P9-AFB-904

The Machine in the Nursery

The Machine in the Nursery

Incubator Technology and the Origins of Newborn Intensive Care

Jeffrey P. Baker

THE JOHNS HOPKINS UNIVERSITY PRESS BALTIMORE AND LONDON

Printed in the United States of America
on acid-free paper

05 04 03 02 01 00 99 98 97 96 5 4 3 2 1

The Johns Hopkins University Press
2715 North Charles Street
Baltimore, Maryland 21218-4319
The Johns Hopkins Press Ltd., London

Library of Congress Cataloging-in-Publication
Data will be found at the end of this book.
A catalog record for this book is available from the
British Library.

ISBN 0-8018-5173-4

CONTENTS

ILLUSTRATIONS

ACKNOWLEDGMENTS

In reflecting over the course of my research, I am struck by my good fortune in having met some extraordinary individuals from many walks of life who have all contributed to this project in one form or another. The catalyst for this study was a retrospective on neonatology given by William A. Silverman at the 1988 Society for Pediatric Research meeting. Silverman has been one of a number of professional neonatologists with a serious interest in history who have encouraged my pursuit of the origins of their specialty. I would particularly like to acknowledge the assistance and suggestions of L. Joseph Butterfield, Lawrence Gartner, and Paul Toubas. As the successive chairpersons of pediatrics at Duke University, Samuel Katz and Michael Frank have made possible my dual vocations in clinical medicine and history.

While my interest in the history of neonatology may have ultimately derived from clinical experiences, the scope of this project has shifted many times through my contact with individuals I met during the course of my graduate studies and research at Duke University. I am especially indebted to Seymour Mauskopf and Alex Roland for encouraging me to integrate the history of medicine with that of technology. Both have been superb role models as scholars and individuals. I. B. Holley, Kristin Neuschel, Michael McVaugh, Anne Firor Scott, and Harmon Smith have provided helpful criticisms at various stages of my writing. Allan M. Brandt played a critical role during my early efforts to formulate this project and has remained a most valued source of encouragement since. Peter C. English has had the patience to bear with this enterprise from the very beginning; indeed, it was largely through his example that I decided to embark on a career as a physician-historian. I am grateful for his sustained interest and good sense. Robert Brugger and the courteous staff of the Johns Hopkins University Press were consistently helpful and supportive.

From a practical standpoint, this book would never have been completed without the support of a National Library of Medicine publication grant. I am particularly indebted to my program officer, Richard West, for his confidence in my project and flexibility in supporting it. It is with real pleasure that I would also like to express my gratitude to the Josiah Charles Trent Memorial Foundation, which made possible those portions of my research involving travel and archival investigation.

My work among library and manuscript collections was rendered far more pleasurable by the professionalism of the staffs involved. Suzanne Porter at the Trent Collection of the Duke University Medical Center Library, Sue

Sacharski at the Northwestern Memorial Hospital Archives, Richard Popper at the Regenstein Library of the University of Chicago, William Creech at the National Archives, and the staff of the National Library of Medicine were most courteous and helpful. In Paris, the staffs of the archives and special collections of Assistance Publique and the Académie de Médecine were equally generous with their assistance. Paul Toubas was kind to provide me access to his valuable personal collection of papers from the French obstetrician Pierre Budin.

Finally, I would like to dedicate this work to my wife, Alicia, whose support and companionship kept me afloat throughout the long course of my research and writing, and to my daughters Madeline and Rachel, whose arrivals helped bring it all to a close.

The Machine in the Nursery

Medicine and the Historiography of Technology

To an unprecedented extent, physicians today use machines. Technology has become so pervasive within medicine that it often seems to be a force beyond our control. Physicians and their critics may disagree over whether its impact has been positive or negative, but both often display a common understanding of technology as an independent agent driving change. We read of the "technological imperative," the idea that any technology that can be used must be used. Expensive technologies such as computerized tomography (CT) scans and dialysis are cited as inflating the costs of medicine. Doctors are portrayed as employing laboratory tests to the exclusion of the patient's narrative and physical exam. From intensive care to artificial hearts, complex new technologies seem to have evolved faster than our wisdom to manage them, spurring a concurrent explosion of popular interest in medical ethics.[1]

Historians of technology have rarely addressed medicine specifically, yet they have certainly dealt with analogous issues. They will recognize the preceding language as masking a form of technological determinism, the conviction that technology itself can literally determine social change. Since the time of Karl Marx, technological determinists have made their case with all manner of sophistication; perhaps one of their most renowned examples is Lynn White's argument that the stirrup catalyzed feudalism.[2] But in recent years historians examining the relation between technology and society have reversed their former emphasis and have attempted to demonstrate how social relations influence technology.[3] That technology can be influenced by its social context seems hard to deny. The difficulty comes in trying to describe how social relations shape, construct, or determine technological change. Answering this question is not without importance, since it suggests that society may find ways to control new technology rather than simply accommodating to it.

Social scientists in recent years have introduced a variety of models to define this relationship; many of these efforts are often subsumed under the general heading of the "social construction" of technology.[4] Although an intimidating battery of technical terms characterizes much of this literature, some provocative insights have emerged as well. Scholars such as Wiebe Bijker and Trevor Pinch have drawn attention to the interpretive flexibility of artifacts, referring to the fact that different social groups may conceptualize the same device in very different ways. Consensus, in a similar manner, becomes a process of closure; it represents the success of one group in imposing its concept of technology on the others.[5] When a new technology diverges into many forms that are rapidly pared down to a few survivors, the process of technological change invites comparison with natural selection, as George Basalla argues.[6] These kinds of explanations are attractive to those who would seek to divest technology of its connections with science and rationality and call attention to the power relations that influence its development. They may be criticized, however, for sometimes introducing specialized terms without clearly making a case for their usefulness.[7] At times, moreover, social constructionists focus so narrowly on external social influences as to exclude internal factors of greater importance. The influence of the social context may, for example, recede with the development of formal evaluation methods.[8]

In comparison, most historians of medical technology have lived in a veritable Eden of lucid expression. Yet their territory remains remarkably unexplored. Stanley Reiser has written the most ambitious historical analysis of the impact of technology on medicine presently available. He sees diagnostic instruments as responsible for encouraging professional specialization, the centralization of medical care, and the decline of personal interaction between physicians and patients.[9] Yet although he continually highlights the role of technology in changing medicine, he does not clearly explain why new technologies are accepted. His book, in fact, is full of citations from physicians who opposed new technologies. Reiser's work has nonetheless helped to generate further research from other historians on the interaction between technology and medicine.[10] Joel Howell has been particularly notable in this respect. His studies of the X ray and the EKG argue that these technologies did not become popular until an institutional and social context existed to promote them.[11]

The present study began as an effort to extend this work into the realm of therapeutic technology. Given the scarcity of historical work on medical technology, I have chosen a case study format. The device in question is the infant incubator, an apparatus developed in the 1880s to preserve the lives of premature infants. It is of interest in two respects. First, it became an integral part of the first premature-infant nurseries, institutions that later evolved into the newborn intensive care units so prominent in today's hospitals. These institutions have become an oft-cited example of runaway technology; their critics have charged that physicians are engaged in a relentless quest to rescue ever smaller premature infants.[12] Second, the incubator is illuminating as an example of technology transfer. Study of the incubator, invented in France and then exported to the United States, allows the historian to examine how the same technology evolved in two different cultures. Such a comparison makes explicit the assumptions embodied in early American neonatal medicine.

My immediate task is to examine the transformation of the incubator as it moved from France to the United States between 1880 and 1922. Though the primary focus is on the United States, I also examine the ways in which American physicians imitated, and departed from, their French counterparts. The time frame is bounded by two important events in the history of neonatal medicine. The choice of 1880 is fairly clear-cut, since it was the year of the incubator's invention by the French obstetrician Stéphane Tarnier. The end point for the comparison is less straightforward. Although the French retained a reputation as leaders in the field of neonatal medicine until the death of Tarnier's successor, Pierre Budin, in 1907, over a dozen years remained before it could really be said that the Americans had developed their own indigenous tradition. The story in the United States is therefore followed until 1922, the year in which Julius Hess, the figure traditionally considered to be the father of American neonatology, came to prominence through two accomplishments: the publication of the first American textbook on premature infants and the creation of the first permanent American premature-infant nursery.

The most interesting question posed by the incubator during these years is not why it took so long to cross the Atlantic but how it was transformed in the process. The incubator encountered a succession of new social groups and cultures as it left the nest of the Paris maternity hospitals and made its way to a new professional home in the United States. It changed shape and form in the process, acquiring new mechanisms and even new functions. And its meaning changed as well, making it a focus of controversy. In the wake of French claims that the incubator could lower premature infant mortality by nearly 50 percent, the device became a symbol of the promise of modern medicine and technology. So great was this excitement that public fairs and expositions began promoting so-called incubator baby shows complete with live infants. Yet a backlash soon followed, construing the device as a symbol of well-intentioned medicine gone awry, encroaching on a realm better left to mothers than doctors. The incubator became a lightening rod for these anxieties and as such offers the historian a revealing glimpse into the relations of physicians, mothers, infants, and machines at the turn of the century.

The story of the incubator's evolution during these years is a complex one. Nonetheless, it is possible to make some generalizations characterizing the French and American approaches to premature infant care, in spite of the underlying diversity. First, France fostered a maternal approach highlighting the role of the mother in saving a premature infant, while the United States promoted a nutritional and environmental approach. The former conception encouraged simple incubators designed to promote the mother's involvement; the latter, more complex devices to maintain a comprehensive artificial environment addressing many of the infant's needs beyond warmth. Second, these different approaches must be understood in terms of the social contexts operating in France and the United States. In particular, four sets of factors were of critical importance. Three of these—national style, professional identification, and institutional context—were somewhat interdependent. The fourth—gender—seems to have acted more as an independent variable.

The overall argument may be summarized briefly. France and the United States cultivated different professional and institutional arrangements for the care of the premature newborn, dominated by obstetricians based in maternity hospitals in France, and pediatricians in infant hospitals in the United States. The French maternity hospitals offered some critical advantages in raising premature babies, most notably immediate access to the infant as well as the possibility of breast-feeding. American pediatricians in infant hospitals lacked these advantages and thus worked in a setting where premature infant care was inherently more difficult. But these different social contexts did not simply influence the relative success of the particular physicians who tried to care for premature infants; they actually shaped how these infants were perceived. In the nineteenth century, the premature infant, as a distinct entity, did not exist. Rather, mothers and physicians tended to conceptualize infants born before term on a continuum with other weak and feeble newborns. All suffered from a generalized lack of vitality. This ambiguous status persisted after the incubator's introduction and, indeed, led to different perspectives among obstetricians and pediatricians. The former were more likely to see the acute complications of prematurity and appreciate the importance of warmth; the latter generally saw a more chronic population that seemed to require nutrition and general environmental care more than any specific therapy such as heat. Social context thus shaped how medical professionals perceived the premature infant, which, in turn, shaped the evolution of the incubator.

The other dynamic to be explored in examining the incubator's history is that of gender, for in their bid to assume responsibility for the premature newborn, obstetricians and pediatricians were expanding into a realm traditionally occupied by mothers and their nurses. Women clearly did not play a passive role in this process; it was they who decided whether to call a physician or to commit their infants to the hospital. This decision had profound implications for physicians, who were frustrated by what they perceived as the negligence of many mothers in failing to bring their infants to the hospital until their condition was nearly hopeless. Unfortunately, the class of women most likely to use a hospital in this period rarely left behind written documentation, so their motivations are not clear. The indirect testimony of physicians, however, suggests that an enormous perceptual division often separated doctors and mothers regarding the treatment of prematurely born infants. In this context, the incubator itself operated as a symbol of medical expansion into the realm of the mother and infant. It raised the question of whether an infant should be entrusted to a machine or to a woman and whether it should be cared for in the home or in the hospital.

This study concentrates on the testimony of physicians, a necessity imposed by the difficulty of obtaining pertinent evidence from mothers who hospitalized their infants before such practice became routine. But it is hoped that historians of women may find material here that deserves further research. Thanks to the work of Judith Leavitt and others, we know far more now about the history of childbirth from the perspective of women than ever before.[13] We need to understand how the medical problems of the newborn fit into this story and how mothers came to feel that they had lost control over this aspect of their lives, as well. For the story to be examined here ultimately points toward the dilemmas of the present-day intensive care nursery and the continuing struggle between physicians and parents to define their responsibilities for the newborn infant.

The narrative is organized around the trajectory described by the incubator itself as it encountered various social groups and contexts in its journey from France to the United States. It begins by considering how premature infants were understood in the nineteenth century. The story proper then moves to France, examining in sequence the invention of the incubator in the Paris maternity hospitals, the backlash that followed the creation of the first premature-infant nurseries, and the reconceptualization of neonatal care that ensued. We next follow the incubator as it traced a course across the Atlantic to the United States, assisted by inventors and promoted by world's fairs and expositions. Afterward, the narrative studies the device's encounters with the obstetric and pediatric professions in the United States. It concludes by examining the eclipse of the incubator during the turbulent infant mortality campaign that reached a climax at the time of World War I, a period that marked the end of what might be fairly designated the first era of neonatal medicine.

CHAPTER ONE

Between Fetus and Weakling

The history of neonatal care cannot be separated from the history of the recognition and treatment of the premature infant. This point deserves emphasis, given that so much of the recent dispute over newborn intensive care has centered on another category of patients, congenitally handicapped infants. In the past twenty years, the newborn nursery has become, as historian David Rothman has aptly observed, a battleground between two forces unleashed during the 1970s: parental rights and the rights of the handicapped.[1] Controversy has thus centered on such scenarios as that posed by the celebrated Baby Doe case of the 1980s, in which the parents of an infant with Down's syndrome elected with their physician to allow the infant to die rather than undergo corrective surgery.[2] Yet the classic prototypes of the handicapped newborn, Down's syndrome and spina bifida, have never accounted for a significant percentage of the newborn intensive care unit (NICU) population. Premature infants have always constituted the great majority of babies treated in these institutions and have also provided most of the ethical and policy problems encountered in actual practice.[3]

A historical account of the origins of newborn technology, therefore, might logically begin by considering the medical problem to which it was a response; that of the premature infant. Unfortunately, this approach leads to trouble almost immediately. Physicians did not consider premature infants a distinct category of patient before the late nineteenth century, much less a medical problem that demanded a solution. The "discovery" and the treatment of the premature infant arose in parallel, not in sequence. Although the incubator emerged as the first technological response to prematurity, it also helped to define what made a premature infant a distinct medical problem in the first place. The story of the incubator thus cannot be told apart from the history of the emergence of prematurity from the shadow land between fetus and the viable newborn.

In order to clarify this point, it may be useful to examine briefly the current medical understanding of prematurity in the context of neonatal intensive care and then address the problems involved in reflecting this perception back into the past. Even today there remains a problem of definition. Should a premature infant be defined on the basis of shortened gestation or of small size? In statistical studies, the second strategy is often easier, since birth weight is more readily available and more precisely quantifiable than is length of gestation. Much of the public health literature thus speaks of "low-birth-weight" infants, usually defined as being under twenty-five hundred grams. With regard to predicting the course of the individual patient, however, defining prematurity in terms of length of gestation (less than thirty-seven weeks) makes far more physiological sense. An infant's small size at birth may signify factors other than prematurity, such as a genetic syndrome, congenital infection, maternal alcohol use, or inadequate nutrition. Neonatologists thus regularly try to categorize premature infants as either preterm or small for gestational age (SGA), with many babies being a combination of both.[4] Assuming a given newborn's small size reflects premature birth rather than being SGA, its outlook will very much depend on the length of its gestation.

At present, about 6 percent of all infants in the United States are born sufficiently premature to require some kind of specialized hospital care. Most of these, however, are born after at least thirty-four weeks' gestation and are treated in intermediate or special-care nurseries.[5] The most common technologies here, including incubators, intermittent tube feedings, intravenous fluids, and low levels of oxygen, are fairly simple and yet have a dramatic impact on the mortality of mildly premature infants. Tube feedings (known as gavage) help compensate for the frequent poor nippling found among such infants, and intravenous fluids address common physiological abnormalities involving calcium, sugar, and electrolyte metabolism. The effect of maintaining temperature deserves special emphasis, since this factor underscores the effectiveness of the incubator. Premature infants are highly vulnerable to exposure, and the precise maintenance of their body temperature has a dramatic impact on their overall mortality rate.[6]

The sophisticated ventilators, monitors, and other technological apparatus characterizing neonatal intensive care units are reserved for a smaller but far sicker population of infants who have frequently been born more than two months early (thirty-one to thirty-two weeks' gestation). These infants face a rocky course, suffering not only from the aforementioned problems of temperature, feeding, and vulnerability to infection but also from inadequate development of the lungs and other organ systems. In particular, babies born this early often cannot produce surfactant, a complex soap-like substance that normally acts to prevent the lungs from spontaneously collapsing from the forces of surface tension. As a consequence, they experience a condition termed hyaline membrane disease or respiratory distress syndrome, which typically lasts several days until abated by the production of new surfactant. Moreover, even those babies of short gestation who escape this syndrome will commonly experience apnea, a pattern of irregular breathing that can lead to life-threatening episodes of cyanosis, in which the infant turns blue from low oxygenation.

Though milder degrees of respiratory distress can be treated with oxygen, almost all infants born earlier than thirty-two weeks' gestation will require mechanical ventilator support to survive. Such infants frequently develop a range of other complications while on the ventilator, many of which have attracted attention only in recent years and hence are not examined in detail here. The vulnerability of many infants born earlier than thirty-two weeks to intracranial hemorrhage deserves mention, since this particular problem can lead to cerebral palsy and mental retardation. It should be noted that expenditures in neonatal technology today are heavily slanted toward a relatively small number of very premature infants. The simpler technologies addressing temperature, feeding, metabolism, and infection control are sufficient for the majority of premature infants.[7]

This description applies to premature infants as they are perceived today from the perspective of contemporary clinical practice and physiology. It is not a "natural history" of prematurity but a portrayal of prematurity set against the context of contemporary neonatal technology. A premature infant born a hundred years ago may have had the same physiology yet "looked" different to contemporaries working within a different social and scientific framework. And the course of such an infant depended very much on where and how it was treated. We might well surmise that almost all of the smallest premature infants would have died within a matter of hours before the availability of intensive care. At the same time, there remained a far larger population of infants born of seven-to-eight-months' gestation whose fate was precarious but not inevitable. Such infants faced a high mortality in the early days and weeks of life, particularly with regard to the risks of exposure, malnutrition, and infection. It was this population of newborn infants that would prove most critical with respect to the incubator and, therefore, to our present inquiry.

In the nineteenth century, the mortality of premature birth was hidden within the high overall mortality of early infancy. Infant mortality in American cities during the 1800s was probably on the order of 15 percent or more, more than ten times its rate today.[8] So much of this mortality was concentrated in the first days and weeks of life that laypersons no doubt took it to some extent as a fact of life rather than a problem capable of solution. Public health reformers and physicians went so far as to formalize the inverse relationship of age and mortality as manifesting a "law of nature" not amenable to medical art.[9] The English health reformer and statistician William Farr rationalized this concept in almost Darwinian terms to explain the fact that "the mortality in the first year of breathing life rapidly increases as we proceed backward from the twelfth, to the third, second, and first month," a trend he went on to extend through the period of premature birth all the way to the massive destruction of spermatozoa necessary "to secure the continuation of the species."[10] Premature infants thus fell on a continuum between the fetus and the term newborn.

To the extent that they addressed premature infants at all, physicians and laypersons alike thus tended to incorporate them into a vague category of "premature and *weak*" infants. Other adjectives might be substituted, including feeble and debilitated, but the overall metaphor remained one of constitutional weakness. Premature infants were envisioned as being immature, in the sense of being small, puny, and of low vitality, rather than strictly in terms of being born early. Indeed, before the twentieth century fetal maturity and full gestation were not necessarily synonymous. A powerful folk tradition, dating back to the Hippocratic corpus, divided premature infants into seven- and eight-month babies and asserted the former to be much more vigorous than the latter. On one level, this belief served a useful social function, particularly since more than one observer noted that a disproportionate number of thriving "seven-month babies" were born to mothers who had been married less than nine months earlier.[11] More important, it suggests that mothers' estimations of the lengths of their pregnancies should not be taken too literally before recent times. Premature infants were most likely accorded such status by being small or "puny" and hence grouped with other small and weak newborns.

It is, therefore, not possible to answer how physicians understood premature infants in the nineteenth century, with reference to the parallel population defined as premature in the hospital nursery today. What we can do is to look at how these infants appeared from specific professional perspectives. The historical record provides access not to a "real" phenomenon of premature birth but to the premature infant as refracted through the interpretation of observers. Three important professional vantage points emerged with respect to prematurity prior to the invention of the incubator. The first was that of obstetricians working in the birthing room who had access to the first moments of a premature infant's life. The infant's later care was entrusted to women, who drew upon a second perspective to guide their management, the domestic medical-advice literature. The third perspective, that of pediatric specialists encountering premature infants in hospitals, was almost exclusively limited to Europe yet gave rise to a tradition of pathology that greatly influenced the American medical literature.

The Obstetric Perspective: Asphyxia and Resuscitation

The home birthing room was the first context in which physicians met premature infants. Long before women began to call upon physicians for advice regarding their newborn infants, they sought their assistance with the process of childbirth itself. By the late eighteenth century, women had begun to admit male physicians to the birthing room in the hope of allaying the pain and danger inherent in labor and delivery. Yet, as historian Judith Leavitt has demonstrated, the mere presence of physicians in the setting of home childbirth did not fundamentally disrupt the dynamics of what remained in many ways a female-dominated ritual. Whenever possible, women with sufficient means attempted to assist one another through the course of childbirth, providing emotional support and acting as the mother's advocate with regard to medical intervention.[12] In theory, the physician fulfilled a well-demarcated role limited to the welfare of the mother and thus typically handed the infant over to her attendants once it took its first breath. It was when the newborn failed to do so that the physician's intervention might be required. The first category of obstetric literature on the newborn thus dealt not with prematurity but a category that could include prematurity: the apparently dead or stillborn infant. Indeed, particularly during the formative years of the profession it could be said that obstetricians literally viewed the problems of early infancy from the perspective of the delivery bed.

Resuscitation procedures joined the many other techniques used by obstetricians to distinguish their profession from midwifery. In this sense, they served a function analogous to the forceps, bleeding, and medication techniques that had encouraged women to choose physicians rather than midwives as birth attendants in the first place. Early obstetricians sometimes portrayed midwives as backward and fatalistic, too ready to give up on a questionable newborn. William Smellie, one of the first British physicians to take up obstetrics, described a 1757 delivery in which the midwife had placed a baby born two months premature in the closet after having failed to resuscitate it through the "common methods" of whipping, applying brandy, and holding an onion to the child's nose. Two or three minutes later, a "kind of whimpering noise" from the closet prompted Smellie to ask if a cat was inside, whereupon the nurse produced a tiny but living infant who was afterwards reared "with great difficulty."[13] The anecdote suggests that a folk tradition of resuscitation practices predated the obstetricians' development of their own.

During the last third of the eighteenth century, an increasingly aggressive spirit characterized the obstetric approach to the newborn. In 1765, one of Smellie's colleagues, Benjamin Pugh, recommended that if a newborn infant did not breathe immediately, "Wipe its mouth, and place your Mouth to the Child's, at the same time pinching the Nose . . . inflate the lungs, rubbing it before the Fire: by which Method I have saved many."[14] Pugh and others also promoted the use of a hollow "air pipe" to inflate the lungs via the nostrils.

Later writers employed the quill of a feather for the same purpose, although their brief accounts demonstrated no awareness of the technical difficulty of true laryngeal intubation.[15]

By the 1790s, leading British obstetric textbooks recommended a variety of interventions for apparently stillborn infants, progressing from tactile stimulation and warmth to bleeding and artificial respiration. The physician confronting a baby who failed to breathe spontaneously was instructed to excite crying though slapping the buttocks, tickling the soles of the feet, or dousing the nostrils with a few drops of brandy. Warmth was to be provided by means of either hot water baths or wrapping the baby in flannel and holding it near an open fire. A clyster, or enema, might be administered to evacuate the stool. If the infant appeared livid and plethoric (as could happen if the umbilical cord had been clamped tardily), a few tablespoons of blood might be drained from the cord. Ventilation by mouth or the air pipe was employed as described; by the early 1800s, some writers advocated inflating the lungs through the pipe by means of a syringe or a bag made of elastic gum. Others went so far as to suggest chest compressions, though for the purpose of assisting the lungs rather than the heart.[16]

This surge of interest in the treatment of "apparent death" was not peculiar to obstetrics but emerged as part of a broader current of enthusiasm for resuscitation that characterized the late eighteenth century. The London obstetrician Alexander Hamilton, in discussing the newborn in his 1792 treatise, acknowledged not the medical profession but another organization, the Royal Humane Society of London, for having proved "to the world, that *Apparent Death* happens more often than was hitherto believed."[17] Beginning in the 1760s, philanthropic Humane Societies sprang up in European and North American cities, convinced that the lives of many victims of drowning and other accidents could be saved if trained laypersons were available. They therefore set about the business of popularizing a range of techniques that included rubbing the victim, providing warmth, and ventilating the lungs either with the mouth or the combination of bellows and air pipe. Such techniques developed for drowning, in turn, inspired many of those used by obstetricians for the apparent stillborn.

Most important, these Humane Societies played an active role in the dissemination of such knowledge. The Royal Humane Society of London offered free lectures, trained lay "medical assistants," distributed equipment to designated rescue stations, and handed out thousands of pocket-sized emergency instruction cards. Meanwhile, artists commemorated dramatic rescues, ministers extolled listeners to follow Christ's example in the raising of Lazarus, and processions of adults and children saved by the society offered hymns of thanks as they marched through the streets of London.[18] It was as if the popularizers of the Enlightenment, having banished God to the status of clockmaker, were at last ready to incorporate his most unique power, resurrection, into mankind's growing armamentarium of "useful knowledge."

The medical approach to the newborn at the beginning of the nineteenth century thus revolved around the metaphor of rescue. It represented an acute care orientation limited to the immediate response to the crisis posed by the apparently dead newborn. More to the point, resuscitation accorded well with the heroic emphasis that was so prominent a feature of medicine from the late Enlightenment through the early 1800s. Though physicians, midwives, and patients might dispute which techniques were most effective, they generally agreed on a basically prointerventionist model of medicine. Even in the backwoods of the American South and West, the influential do-it-yourself medical handbook *Gunn's Domestic Medicine* assumed midwives to be capable of inflating the lungs with a syringe attached to a pipe inserted in one nostril.[19] The techniques of artificial respiration and bloodletting embodied an aggressive and even invasive attitude toward the newborn, encouraging the attendant not merely to assist but to manipulate the young organism as if it were a piece of inanimate matter.

To the extent that the less emergent clinical problems of the premature infant appeared at all in the obstetric literature before the 1820s, they generally did so in accounts of partly successful resuscitations. One medical writer, for example, advocated resuscitation but noted with regard to newborns requiring artificial ventilation: "Such children, however, are apt to continue weak and puny for a considerable while afterwards, so that it is sometimes no easy matter to rear them." What they required was "particular care and tenderness in their management," the implication being that their survival required the devotion of a mother or nurse rather than a physician.[20]

Yet were mothers to turn to domestic medical guides for assistance before the 1820s, they would find precious little advice for the care of premature and weak newborns. Indeed, much of the advice literature in this era assumed that infants were born healthy and subsequently weakened through excessive tenderness. Eighteenth-century child hygiene writers popularized a regimen of cold baths and cold air exposure of young infants that owed more to the romantic idealization of the noble savage than to observations of newborn physiology. The belief that the constitutions of infants should be hardened through exposure to cold air and water had been promoted by the moral and educational writings of John Locke and Jean-Jacques Rousseau. Stories of native peoples tolerating ice-cold baths even in winter led both writers to believe that the human body entered the world strong and vigorous, only to be weakened by indulgence and effeminate civilized life. Locke and Rousseau thus prescribed a regimen of baths employing ever colder water to temper the spirit and fortify the constitution.[21]

Popular writers carried these ideas to a wide audience, often elaborating on them in the process. William Cadogan's influential essay on nursing, for example, went so far as to assert that "the first great mistake" of nurses was that "they think a new-born infant cannot be kept too warm." His main point was to condemn swaddling, but the practical result was to promote the widespread practice of cold bathing of newborns.[22] William Buchan's enormously popular *Domestic Medicine* cited Rousseau in prescribing cold baths for young infants, since "the defects of constitution cannot be supplied by medicine."[23] There was a heroic tendency in infant hygiene, matching that of heroic medicine. Conversely, the decline of the latter was accompanied by the tentative beginnings of professional interest in prematurity beyond the first moments of life.

Natural Healing: A Domestic Approach to Premature Infant Care

During the second quarter of the nineteenth century, medical professionals began to show more interest in the management of premature infants beyond the confines of the birthing room. Their writings reflected the widespread assumption that there was no fundamental distinction between premature and other weak newborns. Physicians envisioned the life force of premature and delicate infants alike in terms of the metaphor of the taper of a small candle, easily extinguished unless gently nourished and sustained.[24] A conservative and supportive set of principles of premature-infant care, revolving around the provision of warmth and environmental care, arose around this metaphor of vitality as a delicate flame.[25]

At the same time, physicians continued to assign primary responsibility for the infant's welfare to the mother. Two factors helped to sustain this traditional division of activity in the United States. First, doctors were in no position to expand their direct responsibility for the newborn. Invasive resuscitation techniques came under fire as part of the general antebellum backlash against heroic medicine.[26] Prominent physicians abandoned direct inflation of the lungs, citing experimental studies that such manipulation could lead to their damage or rupture. A gentler approach to newborn resuscitation, relying on the provision of heat, became prevalent, one which a layperson or mother could more easily master.[27] Second, while faith in the physician as the infant's would-be rescuer declined, faith in the mother was on the rise. The nineteenth century witnessed the apogee of the cult of motherhood, an ideology that exalted women's special capacities for compassion, nurture, and healing.

Mothers in the antebellum period thus provided most of the routine medical care for their infants and children. Those who assisted them were more than likely women, sometimes in the guise of family and neighbors, less often as hired domestic infant nurses. The latter, it should be noted, were often older or widowed women hired for routine infant care rather than professionals concerned with the management of illness.[28] As a result, the management of the sick infant remained primarily a concern for the mother. And while many women no doubt continued to rely on the advice of neighbors and relatives, a growing number were turning to written handbooks on domestic medicine and child rearing.[29]

During the antebellum years, physicians and educated laypersons generated an extensive literature addressing the moral and medical issues involved in rearing children. Indeed, in the egalitarian context of Jacksonian America, the line between professional treatise and popular advice manual was poorly demarcated, at best; mothers consulted a variety of sources ranging from home medical guides exemplified by *Gunn's Domestic Medicine* to increasingly sophisticated treatises and even popular textbooks. This domestic medicine literature provides the most accessible testimony regarding the treatment of premature infants in the early nineteenth century. The extent to which the advice in these manuals was actually carried out, to be sure, is difficult to say. Yet their popularity suggests that they did resonate with the culture from which they emerged. While they cannot be used to estimate the prevalence of various child health practices, they do provide insight into the available possibilities.[30]

One of the most respected American medical authorities on maternal and infant health in the antebellum years was the Philadelphia obstetrician William Potts Dewees, who wrote popular treatises on the diseases of children and on midwifery. Dewees asserted in 1825 that the state of apparent death at birth (which he designated with the medical term for fainting, *syncope*) had both an early and a late form. With regard to the latter, he noticed that some babies, particularly those born prematurely, underwent initial improvement only to succumb to a more indolent course marked by a cold and pale appearance, irregular breathing, and, ultimately, death. Convinced that premature infants born after the seventh month of gestation "should never be abandoned, because the powers of life are feeble," Dewees believed that constant warmth and the avoidance of "the fatigue of washing and dressing" were the keys to their survival.[31]

Like other infant-hygiene writers, Dewees couched his advice for rearing premature infants within the framework of nineteenth-century environmental medicine. Its central metaphor, shared by physicians and laypersons alike, was that of health as a state of balance between an individual's constitution and the vagaries of its environment.[32] The focus was not so much on specific diseases as on the factors that promoted or prevented the expression of disease in general. Heredity influenced but did not determine an individual's health; illness required the interaction of a susceptible constitution with a deleterious environment.[33] Hence, health reformers, especially those concerned with infant mortality, saw the management of the environment as the key to preserving health. At the same time, they recognized that all hygienic advice had to be tailored to the peculiarities of the individual patient.

It was on the issue of whether newborn infants were naturally stronger or weaker than adults that child-hygiene writers disagreed. Most authors accepted some version of the "natural law" concept, that young infants were particularly susceptible to an unhealthy environment. Like the young of other animals, the human infant needed special protection and care by its mother, the provider of nutrition, warmth, and nurture. This conception of the newborn as inherently delicate, however, required overcoming the persistent eighteenth-century legacy that extolled the infant as naturally robust, requiring "hardening" rather than tenderness. Numerous popular treatises continued to promote this philosophy well into the early 1800s. For example, in 1820 the anonymous female author of *The Maternal Physician,* an early American infant hygiene guide that encouraged mothers that "you may yourself be your child's best physician," reported how she had washed her own infant with cold water in the middle of an inclement winter. Though adding that "many good women have called me cruel, and protested that it was unnatural thus to deluge a poor little innocent with cold water," she nonetheless pointed to her children's health as testimony that the practice "invigorates the system" and "renders the babe strong and healthy."[34]

The main controversy over the care of premature and feeble infants in the early nineteenth century thus centered on the value of the traditional cold bath. It is hard to determine how widespread this practice actually was, much less its impact on the mortality of premature infants. What can be stated is that physicians and medical writers after the 1820s increasingly voiced their agreement with Dwees's belief that the custom frequently destroyed children born feeble. In the words of *The American Lady's Medical Pocket-Book,* "This preposterous and cruel system of hardening, as it is termed, is in fact nothing else than an experiment to see how much an infant can bear without being injured or destroyed."[35] Such condemnations became almost a regular feature of infant hygiene guides after 1830.[36] Medical sectarians were sometimes more sympathetic to the hardening philosophy, a circumstance suggesting that interprofessional rivalry provided some of the fuel for the controversy.[37] The hydropaths, as might be expected, remained among the last defenders of the practice; one such writer asserted in 1852 that he knew of delicate young infants bathed daily in New York with "water as cold as could be obtained" without apparent harm.[38] By this time, however, infant hygiene writers were beginning to invoke experimental physiology to justify their conclusions.

The most quoted authority with regard to the temperature of the newborn was William Frédéric Edwards, a son of an English planter in Jamaica who subsequently studied physiology in Paris under the renowned François Magendie. Edwards's most significant work, which helped propel him to the French Academy of Sciences, was a series of painstaking studies on the temperature control of newborn animals presented in his *De l'influence des agents physiques sur la vie,* published in 1824 and translated into English by British physician and health reformer Thomas Hodgkin in 1832. A vitalist seeking to demonstrate how physical forces such as heat, light, and electricity modified living processes, Edwards sought to determine the range of temperature compatible with survival in various animal species.[39] Introducing his section on how temperature tolerance varied at different stages of life, he wrote:

> Instinct leads mothers to keep their infants warm. Philosophers by more or less specious reasoning, have, at different times and in different countries, induced them to abandon this guide, by persuading them that external cold would fortify the constitution of their children as it does those of adults. We will examine this question by the test of experience, in order to be governed by the observation of nature, rather than the varying opinions of men.[40]

Edwards demonstrated that certain mammals, those born with their eyes closed rather than open, lost heat almost as readily as did cold-blooded animals. From this result he concluded that the same principle applied to premature infants until reaching sufficient maturity to open their eyes.[41] Although Edwards himself confirmed this hypothesis by measuring the temperature of only a few human newborns, his conclusions received further support when two other French investigators, Henri Milne-Edwards and Louis René Villermé, demonstrated that mortality in the period shortly following birth was higher in winter than in summer.[42] They attributed the discrepancy to the French legal requirement compelling parents to register their newborn infants at the mayor's office even in cold and inclement weather. Infant hygiene writers continued to quote Edwards, Milne-Edwards, and Villermé throughout the century on the issue of cold bathing.

Edwards's affirmation of maternal instinct in keeping babies warm also accorded well with the assumptions of nineteenth-century pietism. Many writers used his example as an "argument from design" affirming the wisdom of the Creator, much as they extolled breast milk as the ideal food for infants. The immensely popular Scottish infant hygiene writer Andrew Combe wrote that, by rendering the newborn so completely vulnerable, Nature consigned it to rely upon "the strongest feeling which woman can experience—that devoted love of offspring which seldom fails even amid the agonies of death."[43] Along with many other writers on child health, Combe inspired his readers by elevating the rather tedious subject matter of feeding, bathing, and routine infant care to the level of religious truth.[44] By confirming what Edwards called maternal "instinct," the new research demonstrated the compatibility of science with the conceptions of Christian charity, sacrifice, and nurture that found their ultimate expression in the idealized mother. Writing for a pre-Darwinian audience still cherishing the idea of the harmony of science and religion, infant hygiene writers presented scientific knowledge as the unfolding of the Natural Law. Infant mortality would be lowered by bringing human life into accordance with the principles set by the Creator, discovered by science, and directed by natural theology.[45]

By 1850, physicians and infant hygiene writers addressing the newborn had assembled a domestic approach to premature infant care. More precisely, they had developed an approach envisioning the premature infant as part of a continuum with other weak newborns. The key to raising such infants lay in the provisions of their environment: breast milk, clean and salubrious surroundings, avoidance of cold bathing, and, above all, a loving and nurturing mother. The provision of warmth was an important part of their care but did not require the assistance of a physician, much less a special technology. Domestic means of providing warmth sufficed; the most common recommendation was to place the infant in a padded basket lined with hot water bottles.[46] All of these techniques survived to some degree into the early twentieth century.

The still older eighteenth-century approach of hardening and cold baths had an important legacy as well. Though the more dramatic recommendation of submerging infants in cold water lost favor, the therapeutic qualities of fresh (and particularly cold) air would still intrigue physicians well beyond the nineteenth century. Fresh air became one of the cornerstones of therapy for tuberculosis, another disease process characterized to a great extent as a state of weakness. The resurgence of this therapy in the early 1900s would, in turn, impact on the treatment of prematurity, which continued to be characterized as a state of deficient vitality, even after the laboratory and autopsy room had transformed much of the rest of medical practice. Why the rise of scientific medicine, and of pathology in particular, undermined rather than reinforced the budding hygienic tradition of early-nineteenth-century premature infant care constitutes the last of the three major perspectives to be explored in this chapter.

Scientific Medicine: The Premature Infant as Weakling

To the extent that the domestic approach to the premature infant depended ultimately on the metaphor of weakness, it made little distinction between premature and other small or feeble newborns. Conceivably, this assumption could have changed in the course of the great theoretical transformations that overtook medicine during the nineteenth century. The rise of pathological anatomy in postrevolutionary France brought a new way of thinking about the body into professional consciousness, one that directed the physician's gaze away from the holistic language of traditional medicine and toward internal organs and tissues as the seats of specific diseases. One reason we think of premature infants today as distinct from other newborns is that they experience a set of particular disease processes rarely found in term infants. Foremost among these is hyaline membrane disease, the pathological condition associated with respiratory distress in the premature infant. Even without being examined under a microscope, the autopsy of an infant succumbing to this condition will reveal a failure of the lungs to have expanded normally.[47] It thus seems possible that when French and other European physicians conducted postmortem examinations of such infants, they might have distinguished them from other weak and feeble babies.

The problem with this scenario is that it ignores the social context in which pediatric pathology originated, the European foundling hospital. Charles Michel Billard, the most renowned French pediatric pathologist of the early nineteenth century, conducted his autopsy studies in the Hôpital des Enfants Trouvés, Paris's hospital for abandoned infants. Infant abandonment took place on a remarkable scale in Paris during the late eighteenth century, driven by rising rates of illegitimacy associated with poverty in both the city and countryside. During the final two decades of the ancien régime, Parisian parents disposed in this fashion of roughly forty-five hundred infants a year, over a fifth of all recorded baptisms.[48] These infants, most of whom arrived at the hospital in their first days of life, experienced high mortality rates that were often attributed to "weakness" upon admission.[49]

In such a setting, mild prematurity was difficult to distinguish against the background of other tiny newborns who were born small or who had rapidly lost weight after birth. Billard had only minimal knowledge of the infant's history prior to admission and rarely knew the length of its gestation. He thus attempted to define an entity that corresponded to weakness in early infancy rather than prematurity in particular. Although he failed to demonstrate any characteristic set of lesions unifying such infants, Billard nonetheless coined a term in the 1820s, *faiblesse de naissance* (feebleness of birth), that remained in use, with various modifications, for the rest of the century.[50] Rephrased in later years as *congenital feebleness* or *congenital debility*, the term reinforced the traditional metaphor of prematurity as weakness.

Similarly, in the hospital context it was difficult to associate any pathological findings of particular organs with diseases of prematurity. The failure of physicians to identify any characteristic pulmonary lesions on autopsy illustrates the point. In the 1830s, the German physician Edward Jörg recognized a fatal syndrome in which newborn infants were unable to expand their lungs. He called the condition "congenital atelectasis," employing a medical term synonymous with partial or complete lung collapse.[51] Acquired atelectasis occurred in conditions such as tuberculosis, presumably because the patient was too weak to cough or breathe deeply. Similarly, congenital atelectasis was presumed to be a consequence of weakness in the newborn infant. Premature infants were certainly liable to this degree of weakness and hence might experience the syndrome. But most medical authors saw it as a failure to expand the lungs following birth, caused by any condition resulting in diminished vitality, often in the wake of a partly successful resuscitation. One physician wrote, "An infant so affected looks as if it had been rescued from certain death only to die more slowly."[52] Congenital atelectasis, in other words, came to be regarded as a complication of asphyxia or constitutional weakness rather than of prematurity.

Far from distinguishing the premature infant, the language of pathology thus reinforced its continuity with other weak infants. At best, the advocates of clinical pathology did little to advance the treatment of prematurity; at worse, they actually undercut the hygienic tradition's emphasis on warmth and environment. For after 1850, the management of the child with a weak constitution caused by tuberculosis or other chronic conditions increasingly involved the use of stimulants rather than simply a supportive environment.[53] Contemporary physicians observed the dramatic (though transient) effects of sudden changes of temperature on the appearance of children with other atrophic, or "wasting," diseases such as tuberculosis and advocated a therapeutic regimen of frictions, irritants, and baths.[54] Similar stimulating therapies began to characterize the medical treatment of premature and feeble infants as well. The British pediatrician Charles West, for example, recommended that premature babies be rubbed with camphor, exposed to ammonia spirits, given ipecac to induce deep breathing, and eventually stimulated further by treatment with cinchona and other tonics. While recognizing the importance of a consistently warm environment, he also advocated twice-daily hot water baths as another source of stimulation. West thus used warmth in two different ways, as part of a supportive environment and as a stimulant.[55]

These developments may well have confirmed the fears expressed by a popular writer on infant hygiene in 1852 that "the attention of the medical man is generally so largely occupied with investigations into diseases as the causes of illness and death, that it is apt to pass by the apparently trifling subject of preserving at a due standard . . . the temperature of a new-born infant."[56] British and European pediatric authorities in the second half of the century were indeed more likely to specify the temperature of the bath than of the room. Those that did go so far as to specify the latter provide a reminder of how cool hospital wards could become in the winter, as in the case of one prominent textbook recommending that premature infants be kept in a warm room of at least 59° F.[57] In an age when central heating was neither economical nor widespread, the failure to establish environmental temperature standards for premature infants was far from a trivial matter.

More ominous still, the concept of congenital weakness suggested that the one way in which weakness of the newborn was distinct from that of the older child was in its being congenital. Was an infant weak because it had been born prematurely, or was premature birth Nature's way of expelling a defective or weak fetus? Physicians had always acknowledged that heredity played some role in premature birth but had generally considered it in terms of a predisposition that could be countered by the environment; during the late nineteenth century, however, they spoke of the role of heredity in a more determinist fashion.[58] The environmental optimism underlying the antebellum infant hygiene literature gave way to a more subdued and indeed ambivalent set of expectations regarding weak newborns. In the wake of the *Origin of Species* and the social Darwinism of Herbert Spencer, the value of preserving the lives of such infants no doubt seemed increasingly dubious.[59] Discussions of prematurity were increasingly framed in terms of heredity. The Cincinatti obstetrician William Taylor recognized a broad range of maternal living habits that could predispose to premature birth, ranging from "laborious occupation" to alcoholism, tuberculosis, and what he termed an "enervating life." He went on to single out one of these causes as particularly influential, asserting that "as we all know, syphilis more than all else is the cause of too early birth."[60]

Indeed, congenital syphilis became for American physicians the classic example of premature birth caused by a defective infant.[61] During the middle part of the century, physicians such as Thomas Hutchinson and Paul Diday had begun to enumerate specific clinical signs identifying the syphilitic infant and child.[62] Subsequent investigators expanded the syndrome to encompass constitutional weakness as well as particular lesions, in accordance with the variety of routes that were believed to be involved in its transmission from mother and father to infant.[63] The most influential syphilologist of the late nineteenth century, Alfred Fournier, envisioned congenital syphilis as proceeding along a continuum between the extremes of those infants with identifiable lesions and those suffering from nonspecific debility. He asserted that infants inheriting syphilis from both parents were virtually recognizable by their "native debility": "They come into the world small, singularly weak and puny, poorly developed, wrinkled and shriveled, stunted, with the 'old man look,' as it is usually termed."[64]

French pediatricians created in 1880 a pavilion for syphilitic infants in the hospital of Enfants Assistés (formerly Enfants Trouvés) that would later, as will be seen, be converted into a premature infant nursery. It was equipped with a stable to provide milk from donkeys to avoid any danger to wet nurses.[65] No comparable institution emerged in the United States at the time, perhaps suggesting that anxiety over venereal disease, while quite significant in France, ran still higher across the Atlantic. The *Medical Record* dismissed the French efforts to treat such infants. "After all, why should they not thrive?" its editors jeered; "Would they not have been nursing asses' milk if they had been imbibing from the maternal mamma?"[66]

The early advocates of the incubator after 1880 would look back on the third quarter of the century as a period of nihilism with regard to the care of the premature infant.[67] It is difficult to assess these claims, since so little was written on the subject at the time. Perhaps it is fair to see the principal legacy of the hospital-based, clinicopathological tradition of medicine on future premature infant care as the creation of much of the vocabulary for later discourse. It was a language that in many ways obscured as much as it illuminated, however; prematurity for the most part continued to be envisioned as a state of weakness.

A Question of Viability

Depending on their size, premature infants in the nineteenth century thus appeared before the physician somewhere on a continuum between a fetus and a weak newborn. The smallest were in such danger of immediate death that they might be treated as asphyxiated; the larger ones faced a slower but still substantial pattern of mortality in common with other weak and feeble infants. We have traced how the care of both asphyxiated and weak infants evolved in a pattern that mirrored broader trends in the history of medical practice from the late eighteenth to the mid-nineteenth centuries. Heroic interventionism gave way to conservatism and natural healing and eventually to a sense of pessimism over whether therapy made any difference at all. The management of both asphyxia and congenital weakness followed parallel trajectories, even though at a given point of time the former belonged to the physician's domain of responsibility while the latter pertained to the mother's. To complete this picture, it may be helpful to sketch the position of the premature infant from the standpoint of the American medical profession as of the mid-1880s, at which time reports of the incubator began circulating from France.

Though general practitioners in the birthing room still showed little inclination to manage the problems of premature infants in the home setting, they did manifest a renewed fascination with the very viability of these infants. In sharp contrast to the paucity of medical literature on the management of prematurity, medical journals printed numerous accounts by otherwise obscure physicians recounting the rescue of tiny prematures. A New York practitioner in 1883, for example, recalled an occasion many years earlier of rescuscitating a cold and motionless seven-month infant that had been set aside by its family as stillborn. The infant survived to become an accomplished violin player. "One case thus given back to life and light," the author reminisced, "forms a recollection bright and pleasant upon the thorny path of life."[68] Surviving premature infants could attract considerable public curiosity and perhaps help win a reputation for the responsible physician.[69]

Lending urgency to the viability question was a resurgence of professional interest in pushing the limits of newborn rescuscitation. A rapid succession of new techniques for treating asphyxiated infants appeared in the 1870s. As in the preceding century, some were variants of artificial respiration techniques developed for drowning victims; others were based on anatomical and physiological research.[70] In the late nineteenth century, an asphyxiated infant might be swung overhead by the shoulders (Schultze swingings), folded like an accordian (Byrd's method), or have its tongue grasped by forceps in an effort to stimulate the superior laryngeal nerves (LaBorde's method). The use of alternating hot and cold water baths remained a favorite approach as well. With so many techniques available, and with case reports discussing successful resuscitations requiring forty-five minutes or longer, it was ever more important to define which infants warranted such sustained treatment.[71]

The line between prematurity and stillbirth had important medicolegal ramifications as well. Public concern in the United States and Europe had begun to rise in mid-century over the prevalence of child neglect, abandonment, and even infanticide in large cities. Since discussions of the latter frequently involved newborns, the issue often arose whether the deceased infant had been born living or capable of living in the first place.[72] Legally, no Western country considered infants born under less than 180 days' (or six months') gestation to be viable. Treatises on medical jurisprudence often pushed the line of viability later, suggesting seven months as more realistic.[73] But the issue remained unresolved, encouraging the publication of anecdotal case reports throughout the century.

On a national level, the confusion over drawing the line between fetus and premature infant helped to obscure the contribution of prematurity to infant mortality. Most public health authorities conceded the reporting of stillbirths and early infant deaths to be the weakest component of infant-mortality statistics. Mandatory birth registration arose later in the United States than in European countries. It was limited to a few eastern communities in the late 1800s, and a national birth registration area was not created until 1915. Even at that time, leading physicians and statisticians suspected that many infant deaths occurring soon after birth were either not reported at all or were classified as stillbirths.[74]

The limits of professional interest in the treatment of premature infants beyond the crisis of asphyxia were especially evident in another professional controversy that involved these patients. This was the rise of an intervention that potentially pitted the welfare of the mother against that of her infant, the induction of premature labor. This procedure originated in one sense as an expression of conservative obstetrics. The problem of delivering an infant through an unusually narrow or deformed pelvis had long presented one of the nightmares of obstetric practice, a crisis that often required forceps and in extreme cases the ghastly procedure of craniotomy to dismember and extricate the entrapped fetus. The purpose of premature labor induction was to avoid such a situation by precipitating the infant's birth before it reached full size.[75] By 1870, physicians were finding new applications for the procedure in a wide variety of other conditions that might render carrying a pregnancy to term dangerous to the mother. Premature labor induction became one of the most popular subjects in the obstetric literature. Performed by dilating the cervix with a series of inflatable bags, it posed relatively little risk to the mother; the prognosis for the child, of course, depended on the length of its gestation.[76]

Because premature induction theoretically implied placing the welfare of the mother ahead of that of the infant, it raised moral issues analogous to those of therapeutic abortion. The Catholic Church condemned the practice, which in fact progressed far more rapidly in Britain and the United States than in France. Significantly, Americans defended premature labor induction not because it was safe for the infant but because its danger to the baby was outweighed by its benefits to the mother.[77] The experience of a general practitioner from Maine who reported that only seven children lived out of a series of eleven labors induced prematurely was probably quite representative.[78] Although most of the procedure's advocates advised that the procedure not be performed earlier than necessary, few paid any attention in their review articles to the care of the premature infant that resulted.

Most references to premature infants in the American literature in the years just before the introduction of the incubator, in fact, occurred as brief asides in the context of articles primarily concerned with the induction of premature labor. Nearly all authors affirmed that premature-infant care was best provided by a mother in the home. Their brief directions usually offered no more than a paragraph of generalizations regarding the importance of warmth, nutrition, and "the unremitting watchfulness and zeal of a devoted nurse or mother."[79] Most physicians advocated the use of simple measures for warmth, such as placing the child in cotton near a fire. One leading obstetrician did recommend a double-walled warming tub in 1871 that is discussed later in relation to precursors of the incubator, but his interest was exceptional.[80] Far from spurring medical interest in the problems of prematurity, the induction of premature labor appears to have confirmed the traditional boundaries between physician and mother. Physicians still retained only peripheral authority with regard to a premature infant cared for at home, though their comments did suggest widespread awareness of a few basic hygienic principles. Rearing a premature baby at home required good nutrition and warmth, perhaps provided by wrapping the baby in a padded basket. Most of all, however, it required a devoted mother.

§

By the early 1880s, the premature infant thus remained in an ambiguous position between physician and mother and also between fetus and newborn. If born over two months early, it raised the question for the attending doctor of whether it should be set aside as an inviable fetus or resuscitated in the manner of an asphyxiated newborn. Larger and more-developed premature infants, or those salvaged through resuscitation, were generally turned over to the mother. Medical theory provided two sets of guiding principles for their subsequent management: the provision of a supportive environment and the judicious use of physical and pharmacological stimulants. As a consequence, both heat and cold had their places in therapy, the former to sustain the infant's delicate vitality and the latter to stimulate it. The goal was not to maintain a precise infant temperature but to judiciously balance opposing stimuli in its environment, to nourish the flame of life by alternately sheltering and fanning its flickering taper without blowing it out.

In the United States, however, it is hard to know the extent to which these opposing principles were carried out in practice, since the paucity of medical literature suggests that the care of premature infants remained within the realm of women. Equally difficult to discern is to what extent mothers shared the medical profession's skepticism over the treatment of these infants. All of these issues came into sharper focus with the announcement in 1883 of what appeared to be a remarkably successful technology for treating prematurity, the incubator. The story of its invention carries our narrative to France and to another context not explored in the present chapter, the maternity hospital.

CHAPTER TWO

Tarnier's Invention

For most of the nineteenth century, American physicians took little responsibility for the newborn beyond the first minutes of life. Once the infant had breathed, it graduated from the domain of the physician to that of the mother. Physiological medicine might provide guidance for the treatment of the acute crisis of asphyxia but as yet offered few specifics on what to do afterward for the child born premature or weak. Mothers, frequently aided by nurses or other women, drew from the traditions of hygiene and domestic medicine in caring for such infants. The principles of rearing a premature infant were generally no different from those pertaining to the care of any weak infant. The key elements of such care—warmth, breast milk, and tender nurture—remained shrouded in the mystique of motherhood, despite the scientific language employed by the medical advice literature. The mother's role in imparting or nurturing the infant's vitality after birth represented a continuation of her hereditary influence on the child's constitution before birth. In the home setting, ideal infant care without the mother, or at least a female substitute, appeared inconceivable.

The announcement of a new technology to treat premature and weak newborns abruptly challenged this state of equilibrium in the mid-1880s. At this time, France's leading obstetricians transformed the chicken incubator into an infant warming device known as the *couveuse,* or "brooding hen." The apparatus stood at the center of the first systematic efforts of physicians to single out and treat premature infants. It emerged in a context without any exact parallel in the United States, the Paris maternity hospital. Two names in particular dominated the story of its invention and early development. The renowned French obstetrician Etienne Stéphane Tarnier created the apparatus in the early 1880s and produced statistical evidence of its success in one of the largest maternity hospitals of its day, the Paris Maternité. Pierre Budin, Tarnier's intern and later successor as professor of obstetrics at the Faculty of Medicine, by the 1890s eventually superseded his mentor as advocate of premature infant care.[1]

At first glance, the invention of the incubator appears to be a classic story of a "great idea" that transformed medicine. One of the most influential figures within nineteenth-century French obstetrics, Tarnier shaped the careers

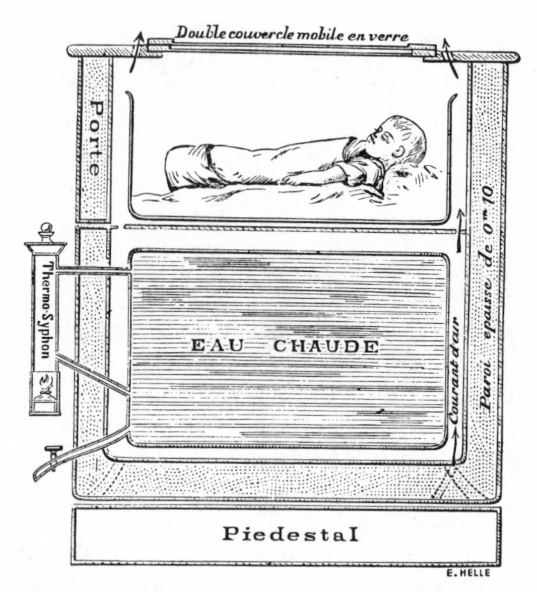

Interior of Tarnier-Martin *couveuse.* Heating source warms water in reservoir via thermosiphon. *Source: American Journal of Obstetrics* 17 (1884): 421.

of an adoring generation of former students who often described his accomplishments in a language more appropriate to hagiography than history. Such accounts understandably explained the invention of the incubator in terms of its creator's genius rather than his social context.[2] After all, Tarnier by 1880 had acquired something of the status of a national hero, a role to which French physicians were more accustomed than were their counterparts in the United States. Stories circulated widely of how he had survived capture by national guards while fulfilling his duties as chief of the Maternité during the violent suppression of the Paris Commune in 1871.[3] His lifelong campaign against childbed fever reinforced his image as a lone crusader against the forces of fatalism and complacency. Tarnier's career, moreover, produced a series of influential inventions such as the forceps that continue to bear his name. The incubator was merely one of his significant innovations.[4]

According to virtually all the early sources, Tarnier conceived the idea of the *couveuse* somewhat serendipitously during a visit to the poultry incubator section of the Paris zoo in 1878. He subsequently had an instrument maker, Odile Martin, construct a similar apparatus in 1880 that by the following year he had begun to use routinely in the Maternité. What was most innovative about the device was its heating system, the *thermosiphon.* Since

Credé's *warmwänne. Source: Archiv für Gynaekologie* 24 (1884): 129.

both alcohol and gas released fumes that precluded their incorporation into the same compartment as the infant, Tarnier required an indirect means of warmth using an outside heating source. Martin accomplished this task by using water as a medium for carrying heat; specifically, an alcohol lamp warmed water in an outside boiler; the water was then circulated through a large reservoir inside the apparatus itself. The device measured close to a cubic meter in size and housed two or more infants, befitting its genealogical connections with multiple-egg chicken incubators.[5] It was an imposing technology that produced dramatic results. With its aid, Tarnier and his staff lowered the mortality of the premature infants under two thousand grams in the Maternité by nearly half.[6] The announcement set off a surge of interest in the device, as rival obstetricians in Paris and later Europe scrambled to develop their own versions of the incubator and to compare them against Tarnier's original.[7]

But just how novel was the *couveuse?* It was certainly not the first hospital technology used to treat premature infants. Although the Maternité had previously relied on the simple expedient of a basket lined with cotton and hot water bottles to warm such infants (who became known by the amusing appellation "little woollies"), a number of other European maternity hospitals had employed special *warmwännen,* or warming tubs, heated by means of a double-walled jacket of warm water. Although the origins of these devices are obscure, the Imperial Foundling Hospital in Moscow and the

Leipzig Maternity Hospital appear to have used them for over twenty years before Tarnier introduced his closed incubator in 1880.[8] Pierre Budin had observed these devices in operation in Moscow in 1878, and Tarnier tried to obtain information on their design from Leipzig soon afterwards. Indeed, in 1884 Carl Credé, the obstetric chief of the Leipzig maternity, asserted that the *warmwänne* had provided the real inspiration behind the French incubator.[9] Credé's recent announcement of the effectiveness of silver nitrate prophylaxis against neonatal ophthalmia had buttressed his own credentials as a leading authority on the newborn infant.[10] His charges consequently set off a priority dispute among the French, Germans, and Russians that is mainly of interest in suggesting that the originality of Tarnier's apparatus was not apparent to all of his contemporaries.

Indeed, the distinctiveness of the Tarnier incubator diminished rapidly as it became simplified during the 1880s. The first model was not only large and cumbersome but provided heat so efficiently that it ran the danger of literally baking its inhabitants. It subsequently evolved in two distinct directions. On the one hand, Tarnier's former intern Budin introduced a mercury thermostat, called a Regnard regulator, activating a battery-powered electric alarm. As the thermostat itself remained a new technology at this time, Budin's contemporaries praised him for developing the incubator into what one called "a true laboratory instrument."[11] But the main thrust of innovation advanced in the opposite direction. The nurses at the Maternité developed a routine of filling the hot water reservoir by hand two or three times a day that struck obstetricians as safer than relying on the thermosiphon. After this modest retreat it was but a short step to reduce the size of the device to enclose only one infant and replace its reservoir with a series of triangular hot water bottles.

The resulting invention, developed by Tarnier and his intern Alfred Auvard, was a model of simplicity. Air entered through small intake holes, passed over a lower compartment with hot water bottles, and circulated upward around the infant by convection before exiting past a spinning anemometer. The physician or nurse could view both the infant and a mounted thermometer through a window on top. The device maintained a constant temperature so long as nurses changed the bottles every two hours, a considerable feat that clearly required close attention. The French instrument makers H. Galante et Fils soon manufactured the Tarnier-Auvard model, which became the most popular version of the incubator until the late 1890s.[12] Further innovation focused on the materials employed, with some physicians preferring metal or glass to the wooden exterior. By the 1890s, Pierre Budin abandoned his thermostatic incubator for the Tarnier-Henry *couveuse*, a wood and glass modification of the Auvard de-

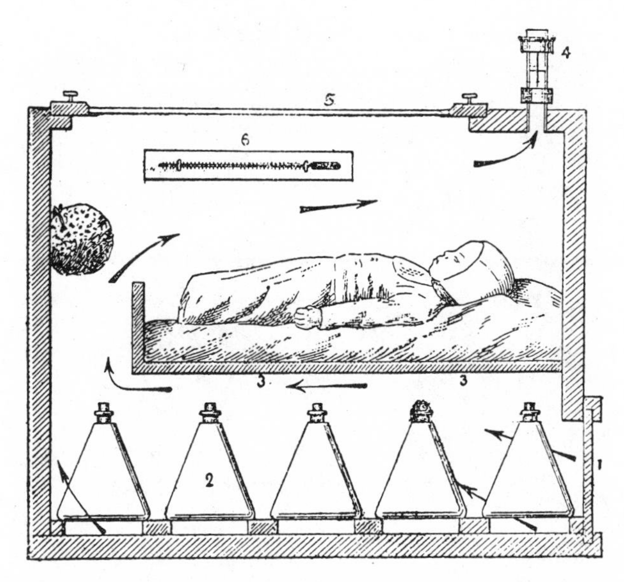

Tarnier-Auvard incubator, cross section. Air warmed by triangular hot water bottles circulates by convection upward around infant prior to exiting via spinning anemometer. *Source:* Pierre Budin, *The Nursling: The Feeding and Hygiene of Premature and Full-Term Infants,* trans. William J. Maloney (London: Caxton Publishing, 1907), 11.

sign developed under the supervision of Tarnier's chief midwife, Madame Henry.[13]

The simplification of the Tarnier incubator as it gained popularity calls into question the extent to which it represented a technological breakthrough. The Tarnier-Auvard model was hardly more complex than a basket with hot water bottles or, for that matter, a double-jacketed warming tub. Its chief distinction was that it was enclosed on all sides. This feature alone commends it to the modern eye as a step forward beyond the padded basket or warming tub, for the simple reason that modern incubators are also enclosed. The rationale behind this design is that a closed incubator reduces heat loss from convection and radiation better than an open model.[14] Tarnier, on the other hand, explained the device with the physiology of his own time; he suggested that it warmed the infant from within as well as without by heating the air it breathed.[15] But the desirability of breathing warm air was a point of controversy in the 1880s, particularly in Britain and the United States. Proponents of the open incubator sometimes argued that the breathing of cold air actually stimulated the infant's lungs, employing the long-held medical view of hot and cold as complementary stimuli, each of

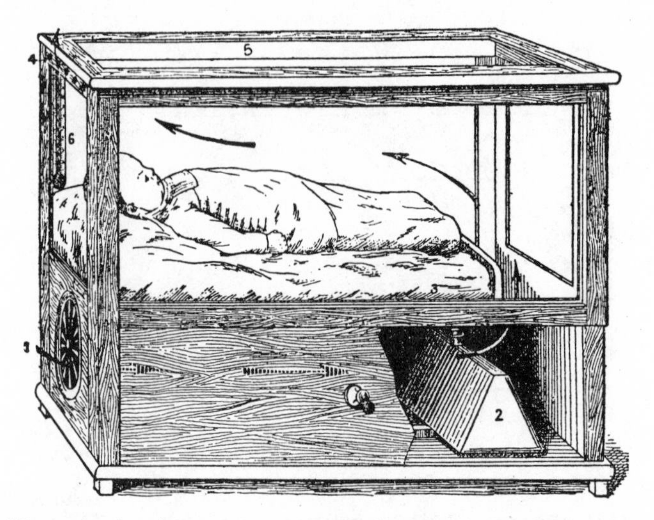

Tarnier-Henry glass incubator. *Source:* Pierre Budin, *The Nursling: The Feeding and Hygiene of Premature and Full-Term Infants,* trans. William J. Maloney (London: Caxton Publishing, 1907, 12.

which required individualized application to a particular case. This qualitative understanding of cold further allowed physicians to reconcile the "hardening" theories so popular early in the century with the recognition of the importance of warmth to weak and premature infants.[16] Even physicians who accepted Tarnier's statistical evidence had trouble understanding why such a simple technology could have such far-reaching results.

Yet it would be equally misleading to agree with Carl Credé that Tarnier, Budin, and their co-workers had accomplished nothing new of significance. The fact remains that Credé did not publish his experience with the *warmwänne* until after the appearance of Tarnier's first article on the incubator. What was most striking about the work of Tarnier and other French obstetricians in the 1880s was not so much a new invention but a new interest in the premature infant. One obstetrician who later recalled Tarnier's work from a more critical standpoint admitted the existence of precedents for the incubator but insightfully singled out Tarnier's special contribution in popularizing the device.[17] Through the introduction of the incubator, the premature infant emerged as a problem relevant to physicians. The new invention may have precipitated this interest, but not in the sense that it appeared as a response to a preconceived problem of how to treat a premature infant. As has been seen in the preceding chapter, the high mortality suffered by premature newborns did not in and of itself render them a problem relevant to physicians. Tarnier's incubator played a critical role in the longer process by which physicians and others defined the premature infant as a clinical entity. Tarnier and his co-workers constructed the concept of the premature infant even as they constructed the incubator.[18]

Before considering the impact of the incubator on newborn care, it is therefore imperative to inquire why the premature infant became a problem for Tarnier and Budin. A biographical consideration of their backgrounds suggests that the premature-infant crusade did not appear de novo, but emerged as a new battlefront out of a broader campaign to transform the maternity hospital from a charitable to a scientific institution.

Origins: French Obstetrics and the Revolt against Nihilism

The career of Stéphane Tarnier is cast against a terrain and time with which few American medical historians are acquainted, that of French clinical medicine in the second half of the nineteenth century.[19] The context is unfamiliar in part because historians have often followed the lead of the American medical profession after the Civil War, for whom Germany replaced France as the most favored destination for study abroad. Historical consideration of France has centered on the critical years after the Revolution, during which medicine underwent a transformation of its own. The most potent symbol of that revolution was the autopsy, the embodiment of a new way of seeing the body in terms of localized diseases afflicting specific organs. A second emblem, especially for Americans influenced by Pierre Louis's "numerical method," was the use of statistics to organize clinical experience. The excitement accompanying both of these methods reached its apogee in 1840, after which the French predilection for bedside examination and the autopsy room over the laboratory became increasingly exaggerated and transformed into objects of scorn in American eyes.[20] Yet the Paris hospitals continued to provide a unique setting for medical investigation, and the characteristic approaches of its practitioners, particularly the numerical method, survived to play a critical role in the story that concerns us here.

The twilight that was supposedly overtaking French medicine in the 1850s made little difference for the institution with which Tarnier was to have such a long association, for the Maternité, even in the high noon of the early 1800s, had never really seen the sun. That was at least the perspective of obstetricians, who during the first three-quarters of the century contended with an institution whose training facilities remained controlled by midwives. The medical faculty possessed one obstetric clinic, to which only students who had passed their third doctoral exam had access, leaving new physicians for the most part devoid of the clinical experience for which Paris hospitals acquired such fame. Instead, midwives employed the hospital to the advantage of their own students, reducing obstetricians to the status of resentful outsiders left to preoccupy themselves over theoretical matters.[21] In many ways the institution functioned more as a hospice than a hospital, a form of institutionalized relief for the poor that offered support and protection for the single mothers who constituted the majority of its clientele.

Although the United States also had a tradition of maternity (or "lying-in") hospitals for the poor and indigent, none of these charities began to approach the scale of the Paris Maternité. The oldest of the major maternity hospitals of continental Europe, the Maternité was opened in the former Abby of Port Royal following the latter's suppression during the Revolution. By 1850, the institution admitted over two thousand women a year. In principle, the city of Paris supervised the hospital's management, but in reality its municipal government showed itself more eager to provide advice than financial support. The combination of the institution's large size and weak financial base set it up for another problem, witnessed on a far grander scale than was ever the case in the United States. In a setting where women could be packed three and even four to a bed, periodic epidemics of puerperal (or childbed) fever could exact a devastating toll.[22]

Tarnier's first encounter with the institution, as a student in 1856, brought him into contact with just such an epidemic. The experience planted the seeds of the most consistent theme of his professional life, the scientific reform of the maternity hospital. The epidemic claimed the lives of 132 of the 2,237 women admitted that year; during the first ten days of May, an astonishing 31 out of 32 consecutively delivered mothers succumbed to the affliction.[23] Tarnier's teachers, he recalled, saw the problem as unavoidable, assuming that any epidemic reigning in the city would strike the hospital. "I revolted," he later remarked, "against the fatalism of such a response, and in the ardor and confidence given by youth, I tried to discover the causes of puerperal fever."[24] Without ignoring the self-serving nature of this recollection, its reference to fatalism still offers an important clue to how Tarnier came to place the experience in the narrative of his own professional life.

The causes of puerperal fever remained a topic of intensifying controversy during the mid-nineteenth century. The most popular explanation attributed its occurrence to the hospital atmosphere. The crowding of patients in a dark and malodorous hospital ward poisoned the air in much the same way as fruit rotted in a hot room. Putrefaction, conceived as a chemical process of degeneration, provided a metaphor for epidemic disease.[25] Tarnier, on the other hand, became through his thesis one of the early proponents of the alternative contagionist explanation.[26] Though already suggested by physicians in various countries, most notably Oliver Wendell Holmes in the United States and Ignaz Semmelweis in Vienna, the notion that an invisible virus or particle constituted the infecting agent of childbed fever had deeply subversive implications for medical practice.[27] Tarnier's accusation that physicians returning from the autopsy room were the agents of their patients' deaths brought into question the reliance on pathological correlation that provided the single most distinctive characteristic of French medicine. The "fatalism" that he described among his teachers may have been no more than a rational medical skepticism quick to scorn hygienic remedies without scientific foundation.

As Tarnier gradually ascended in the Paris obstetric hierarchy during the 1860s, he began to take tentative steps in challenging conventional skepticism regarding premature newborns as well as puerperal fever. During the ten years following his 1858 thesis, he became chief of the Clinique, the teaching center for the obstetric department of the Paris Faculty of Medicine.[28] During this period, Tarnier's interest in the newborn emerged. In 1862, he developed a new dilator to induce premature labor, a procedure that raised the question of what to do with the infant afterward.[29] The Anglo-American practitioners who developed the procedure, it will be recalled, rarely voiced concern over the problems facing the child that resulted. Tarnier, in his annotations to the 1868 edition of Paul Cazeaux's treatise on midwifery, specifically charged physicians "to devote much attention to the child whose imperfect development requires especial care."[30] His concern may have reflected first-hand experience with the raising of premature babies in the hospital, as well as a conciliatory gesture to placate the opposition of the Catholic Church in France to premature induction.

At any rate, Tarnier's specific recommendations remained more conventional than his admonitions. He did not question the standard practice of warming the infant in padded baskets with hot water bottles. In line with most of his contemporaries, Tarnier accepted a room temperature of only 67°F as adequate. Nor did he seem to see the stakes as particularly high. "One does not grieve long over a spontaneous premature delivery," he noted in introducing his discussion of premature birth in 1868, suggesting that the feeble newborn remained more within the domain of nature than of the physician's interventionist capabilities.[31]

The turning point in Tarnier's career occurred following his appointment in 1867 as chief of surgery at the Maternité, after a succession of puerperal fever outbreaks closed the hospital temporarily in the mid-1860s. Following some initial resistance from the hospital administration, he embarked upon a succession of reforms that collectively resulted in the near elimination of the affliction from the hospital wards within the following fifteen years. His actions, it should be emphasized, demonstrated unity of purpose rather than of method. Tarnier set into motion three sets of reforms during the course of the 1870s: the separation of the nursing services attending infectious and healthy patients, the creation of an isolation pavilion with individual rooms, and the institution of Joseph Lister's methods of antisepsis at the end of the decade.[32] Interestingly, Pierre Budin, Tarnier's intern and an enthusiastic advocate of antisepsis after visiting Edinburgh in 1874, specifically noted that Tarnier's interest in the second of the three methods, the pavilion, stemmed from his belief that puerperal fever could be transmitted through the air.[33] In principal, this conviction was distinctly at odds with the other two strategies, which addressed the prevention of infection through contact. Rather than follow a consistent theoretical course, Tarnier chose to try a series of innovations and assess their impact on hospital mortality.

The second prong of Tarnier's method involved statistical assessment of the impact of his disparate innovations. He showed that the maternal mortality rate from puerperal fever fell from 9.3 percent (or up to 20 percent during epidemic years) during the twelve years preceding his arrival in 1867 to 2.3 percent for the twelve years thereafter.[34] The latter figure took into account the successive impacts of separate nursing services, the isolation pavilion, and the beginnings of antisepsis measures; with a longer trial of antiseptic methods in the 1880s, mortality rates fell still further, to only 1 percent.[35] Although later accounts sometimes tried to organize these figures as if they demonstrated the specific impact of a given innovation, such efforts went beyond Tarnier's intent.[36] For Tarnier, statistics served the purpose not of assessing the contribution of specific methods but of showing that his reform program worked as a whole. Indeed, new techniques could always be introduced so long as the statistical trends moved in the right direction. Tarnier's pragmatic method and his use of annual statistics thus reinforced one another, creating, in effect, a powerful engine to sustain and propel further reforms.

These successes converged with the bacteriological discoveries of Louis Pasteur to elevate dramatically the status of French obstetricians in general, and Tarnier in particular. In 1879, Pasteur isolated the *Streptococcus* associated with puerperal fever, placing Tarnier's innovations on a firmer foundation than they had enjoyed at the time of their introduction and pushing him more assuredly in the direction of antisepsis.[37] By 1880, Tarnier had become one of the most admired obstetricians in France, and the idea that high mortality rates necessarily accompanied a maternity hospital had been shattered.[38] Obstetricians, after long years of second-class status, emerged to find their reputations greatly enhanced. One tangible indication of this transformation in status was the establishment by the Paris Municipal Council in 1881 of separate maternity departments in hospitals throughout the city, each to be run by its own obstetric chief.[39] Budin now acquired his own appointment as head of the maternity service of the Charité hospital, though his subsequent collaboration with Tarnier to produce a new textbook demonstrated their continued close relationship.[40] Armed with greater public and institutional support coupled with a renewed confidence, French obstetricians stood poised to expand their domain still further.

Viewed in this context, Tarnier's "discovery" of the premature infant represented the extension of his maternity hospital reform program to encompass the newborn. A sense of momentum had built up in the wake of the optimism generated by the puerperal fever campaign. Unfortunately, Tarnier left no published testimony that directly answered the question of what led him to turn to the premature infant in the late 1870s. According to the account of one associate, he was tormented while supervising the Maternité by the sight of numerous small infants *(petits)* dying on the hospital wards whose bodies became seemingly rigid with cold even though wrapped in fleece.[41] It is not clear that his interest was specifically premature infants at this point, as much as a more vaguely defined population of small and highly vulnerable newborn infants.

As had been the case with the puerperal fever campaign, Tarnier displayed less interest in the etiology of the problem of prematurity than in developing a pragmatic strategy to counter it. The poultry incubator, interestingly, appears to have been a second choice. Most accounts agree that Tarnier did not ask the technician Odile Martin to build his infant incubator until 1880, two years after he first witnessed the device in the Paris zoo in 1878.[42] The reason for the delay appears to have involved Pierre Budin, who at this point returned from a visit to Moscow (undertaken to investigate its antiseptic protocols) with news of the *warmwännen*. Tarnier attempted to obtain a model of the device from Credé's service at Leipzig but apparently gave up on the idea even after the latter sent a design.[43] It is furthermore possible that Tarnier's retreat from the atmospheric model of infection following Pasteur's early discoveries diminished his concern over ventilation and hence over the use of a closed incubator. Though the connections between Tarnier and Credé regarding the incubator are difficult to reconstruct in the wake of the dispute over priorities that followed, the overall story portrays Tarnier as a pragmatist intent more on therapeutic results than providing their rationale.

Again, Tarnier employed statistics to justify the continued expansion of his premature infant campaign, even as he developed other battlefronts besides the incubator. The results appeared in two critical publications in 1883 and 1887. The first and most influential was a lengthy article by Tarnier's intern, Alfred Auvard, published in several Parisian medical journals. Comparing premature infant mortality in the Maternité in the three years before the installation of the device in 1881 to that in the two years following, Auvard demonstrated a decrease for infants less than two thousand grams from 66 percent to 38 percent, a drop of nearly half.[44] The article described both the original Tarnier-Martin model and the simplified Tarnier-Auvard successor, implying (but not clearly stating) that the statistics referred mainly to the former. Tarnier never, in fact, compared one incubator to the other but was content to show through statistics that the concept of the incubator itself worked.

The issue was clouded further as Tarnier began to address the problem of feeding as well as warmth. Finding that the smaller and less developed infants were unable to take breast milk spontaneously, Tarnier introduced in 1884 a small flexible rubber feeding tube designed to pass milk directly to the stomach.[45] He considered this technique, which continues to be known by Tarnier's original designation as gavage feeding, to be second in importance only to the incubator itself. Although Tarnier presented it briefly to the Academy of Medicine in 1885, his student Paul Berthod provided in his 1887 thesis the most detailed statistical analysis of the two innovations yet offered. While still retaining 1881 as the turning-point year, Berthod expanded the pre- and post-treatment groups to encompass five years in either direction, bringing the total number of infants treated with the incubator up to 578. More will be said about this analysis in relation to defining the premature infant; at this point it is sufficient to point out that Berthod stratified his patient population according to the degree of prematurity. His analysis concluded that babies born at seven months or less responded particularly dramatically to the combination of incubator and gavage. The mortality rate of seven-month infants, for example, decreased from 61 percent to 36.3 percent; that of six-and-a-half-month infants fell from 78.5 percent to 47 percent.[46]

As had been the case with the campaign against puerperal fever, the last element to fall into place in the justification of the incubator was theoretical science. In this case, the specific science was physiology rather than bacteriology. Scientific interest regarding the temperature of newborns dated back to the time of William Edwards' experiments in the 1830s but had begun to accelerate markedly during the 1860s and 1870s. These years marked a turning point in medical thermometry most associated with the work of Karl Wunderlich, whose research helped to standardize normal temperatures and promote the use of temperature curves to characterize particular illnesses.[47] Numerous investigators, most of whom were German, attempted to define normal standards and temperature curves for newborn humans and animals.[48] At least one, Alexandre Gueniot, anticipated Tarnier's emphasis on the importance of warming the air breathed by the infant in order to "penetrate" it with heat.[49] But the story of how Tarnier pragmatically developed the *couveuse* only after failing to obtain one of Credé's warming tubs suggests that theory served to justify a course already taken.

This is not to say that the incubator, once invented, encouraged no scientific tradition of its own. Tarnier's students and interns, always on the search for material for theses and studies, found new opportunities to go beyond their teacher. Auvard, for example, not only collaborated with the development of a new incubator model but also conducted painstaking studies describing the temperature curves of premature infants over the first days of life. He monitored the infants as often as every hour during the first day and four times daily thereafter. In the process, he reached an important conclusion: the incubator restored the temperature of a chilled infant to the normal range within six hours, far more rapidly than the four to six days that might otherwise be required for recovery.[50] Nonetheless, Auvard did little to integrate his research on temperature with his therapeutic recommendations. He and the other physicians in the Maternité devoted remarkably little attention in their writings to the question of how warm the incubator should be kept. There was ambiguity over whether to rely on the infant's rectal temperature or that of its head and extremities, which fluctuated more dramatically. Other signs, ranging from the respiratory and heart rates to irritability and color, seemed to be important means of monitoring the incubator's temperature as well. Finally, the infant required different degrees of warmth depending on its size and age.

With so many factors involved, it is perhaps not surprising that Tarnier and his associates emphasized the use of judgment, rather than specific guidelines, in maintaining the incubator at a proper temperature for a particular baby. Though they recommended monitoring the baby's rectal temperature at least twice a day and keeping it close to the adult normal value of 98.6°F, they did not emphasize the importance of precise and frequent measurements in their publications.[51] As Tarnier and his students gained experience, they became comfortable in recommending that the incubator be kept within a temperature range of 86–98.6°F, usually around 90°F.[52] Such a wide range further highlighted the importance of clinical judgment in supervising the incubator and would later raise questions for professionals in the United States and other countries who lacked first-hand experience on Tarnier's service. Ultimately, the statistical rationale for the incubator proved to be easier to communicate than its physiological justification.

It would, therefore, be deceiving to present the incubator's invention simply in terms of technology as applied science. Tarnier justified the device in primarily pragmatic terms based on his experience, bolstered with simple statistics. He left its scientific explanation to his students and interns. Their work at times demonstrated the beginnings of a physiological research program revolving around the problems of prematurity and the incubator. Berthod, in particular, made an intriguing reference to the incubator as an "artificial millieu" in which the effects of various gases, particularly oxygen, on the infant might be studied in the future.[53] The incubator had the potential to become an apparatus for scientific research, a kind of artificial environment in which investigators could examine the response of living systems to nonnatural external conditions.[54] But this approach never became dominant within the Paris obstetric tradition, whose members concerned themselves more with promoting the device than with understanding why it worked.

The incubator thus emerged as an offshoot of an obstetrical program of maternity hospital reform, justified primarily by survival statistics rather than physiology. French obstetricians focused on treating populations rather than studying individuals. In the process, however, they categorized the population in question in new ways. The process of technical invention and refinement was tightly interwoven with that of the discovery and characterization of the premature infant. Both medical problem and technological solution evolved in concert, each shaping the other. Considering how the two interacted helps to illuminate how the incubator did not so much arise in response to a medical problem but rather helped to define what that problem was.

Preterm versus Weak Newborns: Defining the Problem

Before addressing the expansion of the incubator campaign in the 1890s, we need to clarify further how Paris obstetricians regarded the premature infant. Thus far, we have spoken of Tarnier as having discovered the premature infant, at least in the sense of bringing it into the medical sphere. Even momentary reflection, however, suggests that "discover" may not be the best word to apply to a phenomenon that has presumably always been a part of human experience. Just as Tarnier constructed the incubator, he also created a conceptual framework that defined what it meant for a baby to be born weak or premature. How he did so had implications for the response of the government and public to his work.

The best way to understand the distinctiveness of the French perspective is to consider the path its obstetric leaders rejected. German pediatricians during the late 1800s by and large organized their textbooks around the diseases of infancy without distinguishing premature infants as a special population. Edward Henoch, one of the leading pediatric teachers of his generation, incorporated no special section on prematurity in his 1882 textbook. Instead, he included a section on specific diseases of the newborn, such as jaundice, infection, and tetanus.[55] The French obstetricians did in fact attempt to treat a variety of newborn malformations and diseases in the incubator; Auvard's series included occasional infants with cleft lips, spina bifida, fractures, and congenital syphilis.[56] But at an early stage they began to center their attention on small and premature babies. What the French obstetricians did, in other words, was to focus primarily on the problem of the premature and weak infant and then to relegate a limited number of specific diseases to the level of complications of prematurity. They created a two-tiered classification structure of constitutional weakness with a set of particular complications.

Although this primary focus on individual constitution rather than specific diseases was in a sense old-fashioned by the late nineteenth century, the French reinvigorated it by developing a special language to describe the pathology of prematurity. Specifically, they identified the older notion of congenital feebleness or debility, a concept dating back to Charles Michel Billard in the 1820s, much more assuredly with premature birth. Tarnier and his associates used the designation "preterm" *(avant terme)* almost interchangeably in papers and statistical tables during the 1880s with "congenital feebleness" *(faiblesse congénitale)*.[57] No doubt the newer use of the term reflected the fact that French obstetricians, unlike their pediatric colleagues, cared for the mother as well as the infant and hence usually knew whether or not an infant was born early. Even when they used the term for weakling *(débile)*, they generally referred to infants born prematurely.[58]

This primary focus on constitutional debility did not preclude the recognition of specific diseases of prematurity. But it did shape their perception. French obstetricians not only tended to explain most of the complications of prematurity in terms of weakness but described infants' symptoms in ways most compatible with this same orientation. The contrast in the way they described respiratory distress with today's perspective illustrates the point. Many of the premature infants in the Maternité were under two thousand grams and thus should have manifested respiratory symptoms that today would be divided primarily into the two categories of respiratory distress syndrome and apnea. The former syndrome, representing inadequate lung development, causes symptoms of increased respiratory effort: afflicted infants strain the muscles of their thin chest walls to the point of retracting the ribs and emitting weak grunting sounds instead of a healthy cry. But premature infants also commonly experience apnea, effortless cyanotic or blue spells associated with immature neurological control of the respiratory drive.

Nineteenth-century pediatricians in Britain and Germany, who tended to see these infants in terms of the pathological diagnosis of congenital atelectasis (failure to expand the lungs after birth), did appreciate the role of increased respiratory effort.[59] By comparison the French played down the prevalence of respiratory distress in favor of apnea or cyanosis, which they, in turn, attributed to low temperature. Budin, in describing the premature infant, highlighted the primacy of diminished vigor and vitality. The afflicted infant manifested "a body small and thin, limp and transparent, [with] incomplete respiration. The inertia of the external muscles is striking, the cries are without vigor, the sucking motions are feeble and insufficient."[60] Whereas today respiratory symptoms are seen as indications for oxygen therapy, in the 1880s they called for the a boost of "energy" provided by the incubator.[61]

The closest the French came to defining the target population of the *couveuse* in terms of a disease rather than constitutional weakness was with regard to a diagnostic entity almost forgotten today. *Sclerema* was a clinical syndrome seen principally late in the course of hospitalized premature infants and foundlings in which the skin became yellow, hard, and leathery. The condition was usually fatal, terminating in a state resembling rigor mortis, with shallow respirations and ultimately death. Physicians did not understand its etiology in the nineteenth century, though they were aware of its association with prolonged hypothermia in premature and weak infants.[62] At any rate, many of the early incubator patients had varying degrees of this condition and often responded dramatically to the device's application. Berthod asserted that the incubator had a "heroic" effect on sclerema and went so far as to claim that it provided a "quasi-specific" therapy.[63] Indeed, the entity became far less common as incubators became widespread in the 1890s, suggesting at least some connection between the syndrome and hypothermia.[64]

But Berthod's claim to have found a cure for a traditionally lethal disease was to a great extent a matter of redefining the problem. There were questions over the relationship between sclerema and edema, a more common (and far less serious) condition of soft tissue swelling that also afflicted premature babies. Auvard had followed the example of the prominent French pediatrician Jules-Joseph Parrot in distinguishing the two; Berthod grouped them together. He incorporated Auvard's cases of edema into his own case series of sclerema and, not surprisingly, obtained dramatically low mortality rates with the incubator. The example illustrates how disease definitions could cause the efficacy of the incubator to be exaggerated.[65]

The great innovation of the Paris obstetricians with regard to prematurity was not to define a new disease but to use the birth weight as a quantitative standard to categorize congenital feebleness. Initially, there was disagreement over whether premature infants should be classified according to their gestation or their size. The problems with using the former have already been mentioned: mothers tended to describe their pregnancy in terms of months, and even then popular tradition allowed a certain degree of flexibility in embarrassing situations involving illegitimacy.[66] Birth weight was superior for statistical purposes, because it could be measured.[67] It provided a seemingly objective means of characterizing and subdividing the spectrum of prematurity. The stratification of infants according to birth weight also facilitated comparison; critics might argue over whether Berthod had defined his way to a cure for sclerema but had more trouble challenging Auvard's "hard" statistic showing that the mortality of premature infants under two thousand grams fell by nearly half with the use of the incubator. The classification of prematurity according to birth weight was a major accomplishment of the French obstetric tradition, one that would only gradually be appreciated in the United States.

Yet even numbers had to be interpreted, and Tarnier found ways to present his statistics in a way that portrayed his invention as pushing back the limits of viability. The main place of incubators today is in the treatment of mildly premature infants; those born before thirty-one weeks (or around fifteen hundred grams) frequently require some kind of respiratory assistance as well.[68] Tarnier conceivably could have targeted the incubator to those relatively large infants born less than one month early, almost all of whom survived with the device. Instead, he and his associates tended to emphasize the smaller and less mature infants, where the apparatus had the greatest impact. The French obstetricians considered infants between roughly fourteen hundred and two thousand grams to be the best candidates for the incubator; those born still smaller frequently died even with its help, and those larger often survived without the apparatus. And they were constantly interested in testing the lower limits of this range. Auvard cited cases of surviving infants as small as eleven hundred grams.[69] Berthod produced a dramatic table that displayed premature infant survival according to length of pregnancy, in spite of the many questions regarding the accuracy of this method.[70] The results apparently demonstrated that the incubator made possible the survival of some infants down to six months' gestation, an astonishing claim given that such infants face a highly guarded prognosis, even with the assistance of intensive-care technology today.

Such statistics led Tarnier to conclude that he was on the verge of pushing back the medical definition of viability to its legal definition of six months (or 180 days) of gestation.[71] His statement, later to be challenged, exemplified the remarkable optimism that characterized the first years following the introduction of the incubator.[72] There was a sense that a fundamental barrier had been broken, that medicine had essentially redefined the practical limits of viability.

In retrospect, one suspects that the maternity hospital context fostered this spirit of confidence in ways that its participants did not fully appreciate. First, and most obviously, it provided a great advantage in allowing virtually immediate access to the newborn. The incubator could be applied quickly after birth, without allowing time for the infant to deteriorate. Second, breast milk was often available, although in the early 1880s infants too weak to suck were often fed milk from donkeys.[73] Third, and perhaps less obviously, the hospital circumscribed the limits of what premature infant survival meant. The ease with which physicians in training such as Auvard or Berthod were able to collect patient statistics in the hospital made it possible to overlook whether their differential survival rates continued to apply after discharge. In the 1880s, the question of prognosis, the possibility of a handicapped or impaired survivor, did not appear to trouble the confident obstetricians of the Maternité. They continued to be carried along by a technological optimism that flowed naturally out of the previous decade of maternity hospital reform.

Institutional context, therefore, served as one more parameter shaping medical perception of the premature infant in the 1880s. The maternity hospital provided a relatively controlled setting in which premature infants could be meaningfully compared in terms of hospital survival and birth weight. Physicians also controlled much of the overall care of the infant, certainly to a far greater extent than would be the case following a home birth. In the obstetric hospital setting, the incubator indeed appeared to be a therapeutic breakthrough. At the same time, powerful forces were emerging to disrupt this fairly self-contained institutional context, for by the 1890s, Tarnier's premature infant campaign was taking on a momentum of its own and was beginning to move beyond the walls of the maternity hospital. As it did so, it acquired a new public health rationale linked to France's increasing preoccupation with depopulation and national decline.

❧

The incubator was not simply a "cure" for prematurity, or any other disease. Its success very much depended on the choice of the patients placed inside it. Had the device merely been applied in an anecdotal fashion to particular premature babies, it might well have been considered a failure; it was only through large statistical analyses that its relative effect was best demonstrated. Tarnier's greatest contribution was not to produce a technical solution for a preconceived problem but to define the problem in a way that showed it to be amenable to medical intervention. He and his associates did so by taking a heterogenous patient population represented by the terms *premature* and *weakling* and subdividing it according to the quantitative standard of birth weight. The real innovation behind the incubator, in other words, was not in its rather rudimentary technical design but its definition of the target population most likely to benefit from it. It was a therapy that had the greatest impact on premature infants between fourteen hundred and two thousand grams, not necessarily all infants of the same weight.

Secondly, the invention of the incubator can be understood to a great extent in terms of a fairly limited context, the nineteenth-century maternity hospital reform program led by Stéphane Tarnier. Both medical science and wider social factors played only secondary roles in its development throughout the 1880s. This is not to say that the latter factors were entirely absent; newborn temperature physiology did help to justify the incubator, and French society made possible the existence of maternity hospitals. Nonetheless, even without evoking such factors, it is evident that a powerful sense of momentum had been created within the hospital itself by the late 1870s that encouraged Tarnier and others to challenge conventional "fatalistic" notions regarding both maternal and newborn mortality. It was at this point that science came into play, not so much as a source of specific techniques but as a source of inspiration. Having effected a spectacular decline in maternal mortality from infection, obstetricians were poised to capture the newborn as well.

The outside world broke into the confines of the maternity hospital much more forcibly during the 1890s, at which time the story of the incubator must be told from a wider perspective. At this point, the spotlight shifted from the aging Tarnier to Pierre Budin as the most influential advocate for the incubator and premature infant. Budin saw his professional career as a continuation of the themes that had driven his former teacher Tarnier; in a speech to socialist workers in 1893, he integrated the Pasteurian revolution with that of his own profession:

> We have assisted in these final years a veritable revolution. In former times, there reigned in the maternities a very deadly affliction, puerperal fever. Today, thanks to the work that followed the discoveries of Pasteur, women in delivery no longer have to die. . . . We hope that we will assist a similar revolution for infants. . . . All, poor and rich, will be able to conserve their little ones and enjoy the sweet joys of the hearth without experiencing its sadness, all of which promotes the great beneficence of the *Patrie*, which needs all its children.[74]

Budin's final allusion to the imperative of conserving children for France highlighted a factor that would become increasingly important in shaping the future of the incubator, the country's mounting concern over the implications of its low birthrate. Depopulation anxiety, in fact, would push the incubator into a new role going well beyond the walls of the hospital. In the process, it met its first setbacks. Budin's "revolution," like most revolutions, failed to follow the course intended by its instigators. By the end of the decade, the premature infant campaign was under fire, and Pierre Budin was at the center of the controversy that followed.

CHAPTER THREE

Mothers and Nurslings: The French Incubator Campaign

During the early years following the invention of the incubator, the treatment of the premature infant represented more of a medical challenge than a public health imperative. Stéphane Tarnier developed the device in the context of a program to reduce hospital mortality. He was fascinated by the possibilities of challenging conventional wisdom (and fatalism) regarding the outcome of young premature infants, of pushing back the limits of viability. But beneath the optimism of the incubator's honeymoon period during the 1880s lay the unsettling reality that its use had been limited to a context where only a fraction of all infants were born, the maternity hospital. Though hospitalized birth was becoming more common in Paris than in any comparable American city, it remained confined to less than a quarter of all births by the end of the century.[1] The impact of hospital incubator care on overall infant mortality was therefore quite limited.

The incubator entered a new phase of development in the 1890s, as physicians turned to the problem of treating the majority of premature and weak newborns who continued to be born at home. Hospitals began to open special *services des débiles*—literally, departments of weaklings—designed to admit such infants without their mothers. The Maternité opened the first and largest of these establishments in 1893; others followed over the course of the next ten years.[2] All relied on incubators, principally modifications of the simple Tarnier-Auvard design as well as a limited number of more advanced models. They also employed a characteristically French solution to the problem of infant feeding, the wet nurse. In theory, the incubator and the wet nurse could provide the warmth and food necessary to rear a premature infant apart from its mother. The sophisticated level of care in these nurseries was to extend Tarnier's successes from the hospital to the city at large.

But the new institutions created a crisis for the future of incubator care. Physicians found they had no control over the infant before admission. The hospital became an open system with the parent able to influence profoundly the physician's ability to treat the patient. The result was that the tension between the competing roles of physician and mother as caretakers of the infant, always potentially a source of disruption, came out into the open. During the controversy that followed, France developed a distinctive approach to the premature infant, reflecting its contemporary social values and national priorities. Foremost among these was mounting anxiety over the role of the mother in relation to the nation's perceived demographic and political decline. By the 1890s, fears of depopulation were becoming a national obsession, creating a climate that would influence the evolution of the incubator in powerful but conflicting ways.

Depopulation, Decline, and Congenital Debility

Depopulation anxiety was not new to France in the 1890s. It had been gradually increasing in strength since at least the middle of the century. In the time of Napoleon, France had been the most populous country in Europe. Yet it also became the first modern European country to experience a decline in its birth rate, a trend that became evident to demographers by the 1850s. By 1860, France's population had slipped behind that of Austria and within another dozen years behind the newly united state of Germany as well. Indeed, the single event most responsible for catapulting these statistics to national attention was France's embarrassingly swift defeat by the German Empire in the Franco-Prussian War of 1870–71. Public figures blamed the loss on a growing national degeneracy and saddled the nation's mothers with much of the responsibility. Mothers were neglecting their children, it was alleged, weaning them too early or sending them out with wet nurses of dubious character in the countryside. And on an even more fundamental level, they were forsaking their patriotic duty to bear enough children to maintain the future armies of the Third Republic.[3]

But although depopulation also provided a rationale for preserving the lives of infants already born, its use in this regard for the most part awaited the 1890s. The commitment of the Third Republic to social welfare remained tentative in its early years. Conservative politicians showed more interest in offering economic incentives for large families than developing maternal- and infant welfare programs. The 1890s, however, witnessed the emergence of a solidarist reform movement whose leaders envisioned welfare programs as the social glue with which to bind the fragmented elements of French society.[4] Their attitude toward women and depopulation was more complex than that of the conservatives, as exemplified by the prominent physician-politician and infant welfare advocate, Paul Strauss. Though undoubtedly sympathetic to the social problems faced by mothers in raising their infants, Strauss and fellow solidarists still tended to take interest in women for their role in producing and raising children.[5] Such an emphasis had considerable implications for obstetricians.

In fact, the obstetric profession became one of the greatest beneficiaries from the depopulation debate. Its leaders were able to tap into an outlook that not only made the reduction of infant mortality a national interest but saw the mother as the key to its amelioration. French obstetricians were encouraged to move beyond midwifery (the management of labor and delivery) to the more comprehensive management of pregnancy and newborn care. In return, the government acted as a powerful ally, providing new facilities for training and patient care. The twenty years following the invention of the incubator witnessed an unprecedented expansion of hospital obstetric services in Paris. During these years, the annual number of women giving birth in the hospital more than tripled, from 5,180 to 16,299, even though the city's overall birth rate remained constant at around 64,000 annually.[6] New maternity hospitals sprang up around Paris, and by 1893 the law explicitly guaranteed any woman free hospitalization for childbirth. Though poor and unmarried women continued to provide most of the admissions to maternity hospitals, a growing number of married women joined them as well.[7]

The expanding mission of maternity hospitals provided a receptive climate for the incubator while suggesting the need for a more systematic approach to its application. The municipal director of public assistance took action to endow the device in the majority of Paris maternity establishments almost immediately after the first reports of its success.[8] In 1885, members of the prestigious Academy of Medicine packed its amphitheater to hear Tarnier personally discuss his new device. The academy's president singled out the new invention on the grounds that "independent of its scientific interest, it has the merit of timeliness . . . at a moment when the growth of the population is experiencing, in France, a disquieting slowing."[9] Meanwhile, press reports on the incubator fanned popular interest further, with the consequence that mothers began bringing premature infants born at home to the doors of the Maternité for treatment in the *couveuse.*[10] The hospital was hardly ready for the extra patient load.

During the first few years after Tarnier began to use incubators, he and his staff had given little thought to integrating them into the hospital environment. By the late 1880s, six incubators stood in a crowded room next to the quarters of the five wet nurses or *nourrices.* The room was called a *crèche,* a term popularly used to designate charitable day care facilities for working mothers.[11] The term was appropriate, given the lack of interest in the involvement of the infant's mother. Although virtually all the healthy infants at the Maternité were breast-fed, the incubator baby *crèche* was too isolated from the rest of the hospital to make maternal nursing practical. The wet nurses fed as many of the infants as possible, while the remainder were offered milk from donkeys, an expedient previously used by the pediatrician Parrot for infants with congenital syphilis.[12]

Incubators in the Maternité, Paris, 1884. Wet nurses and supervising nurses care for infants in both original and simplified Tarnier incubator designs. *Source: Illustrated London News,* 8 Mar. 1884, 228.

At this point, the chief midwife of the Maternité, Madame Henry, began a campaign to create a special service for premature infants within the hospital. Although midwives were responsible for much of the record keeping that made the statistical studies of Auvard and Berthod possible, only Henry played a more public role in the promotion of the incubator. As increasing numbers of mothers brought their infants from the outside, she proposed to Tarnier that the Maternité create a separate ward to house the incubator babies. Tarnier supported the idea but found the financial aspects of the project daunting. Undeterred, Henry raised one thousand francs by organizing a private charity. She then enlisted the support of Paul Strauss, already at this point a rising member of the Paris municipal council.[13]

Henry's timing was propitious, for in 1891 the national concern over depopulation reached new heights. The census results of that year revealed for the first time that during the previous five years, the number of deaths had actually exceeded the number of births. The specter of an absolute decline in the national population precipitated a sense of near panic among public leaders. The nationalist writer Emile Zola began work on his pro-natalist didactic novel, *Fécondité,* that would be published at the end of the decade.[14] But while conservatives intensified their efforts to promote motherhood, re-

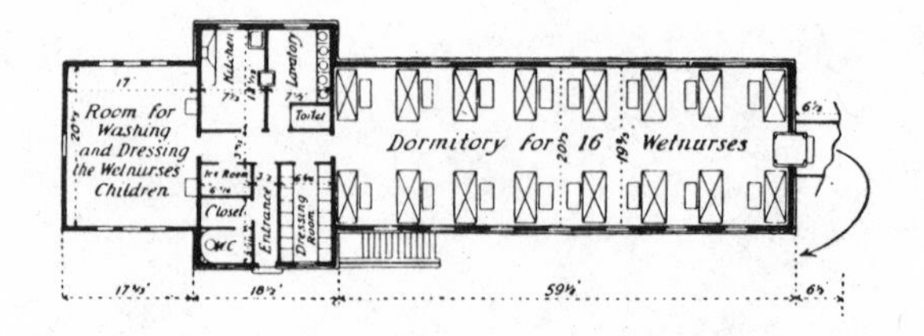

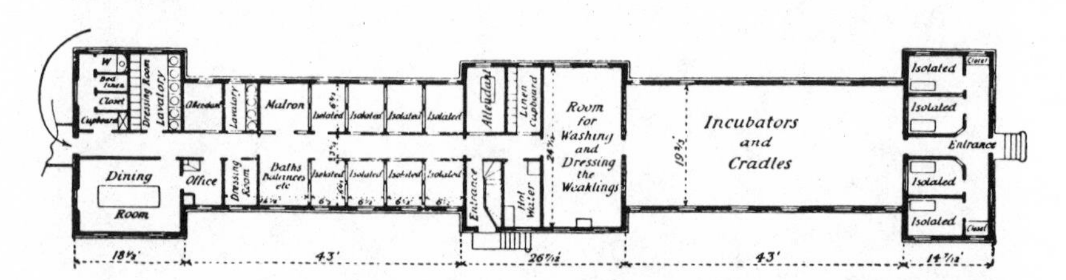

Pavillon de débiles, Paris Maternité, after completion of separate residential area for wet nurses in 1897. *Source:* Pierre Budin, *The Nursling: The Feeding and Hygiene of Premature and Full-Term Infants,* trans. William J. Maloney (London: Caxton Publishing, 1907), 59.

formers found new ammunition for the strategy of preserving infant life. In an 1891 speech, the prominent politician Jules Simon employed a widely used metaphor in proclaiming that infant mortality robbed France of a battalion of future soldiers every year.[15] The decision of the Paris municipal council to fund Henry's service for weaklings in the same year was hardly coincidental, given Strauss's involvement. It approved a total of forty thousand francs in 1891 to allow construction of the new service, later supplemented by an additional seventeen thousand francs prior to its completion two years later.[16]

The *pavillon des débiles* of the Maternité opened in 1893 and immediately became the most extensive special care nursery for premature infants of its day. A spacious room with high ceilings and a forced-steam heating system, the pavilion contained fourteen glass incubators arranged in two rows. An autoclave served to sterilize garments and utensils. Adjoining the ward were living quarters for the pavilion's own staff of wet nurses, enabling infants to be fed apart from their mothers.[17] The pavilion of weaklings symbolized the culmination of the previous decade of caring for premature infants in the maternity hospital setting, as well as a dividing line introducing a new phase of newborn medicine. The institution served a mission extending beyond the Maternité to the far greater population of premature and weak newborns born at home. With its creation, treatment of the premature infant expanded from the professional to the public stage.

It was perhaps fitting at this point that the medical leadership should change as well. In 1895, Pierre Constant Budin took over Tarnier's former position as obstetric chief of the Maternité. Although he had previously played only an ancillary role in the development of the incubator, during the 1890s Budin became the most visible champion of the premature infant. By this point, Tarnier's health was waning as he entered the last years of his life; his time was consumed by professional duties at the Paris Faculty of Medicine and the Academy of Medicine (of which he became president in 1891). Meanwhile, political rivalries conspired to thwart the professional ambitions of Tarnier's former collaborator Alfred Auvard, who never succeeded in obtaining one of the limited number of Paris hospital obstetric appointments.[18] Budin's star rose as Auvard's fell, particularly through his collaboration with Tarnier in his revised four-volume textbook on obstetrics in 1886.[19] Yet although Budin had the reputation of being Tarnier's favorite intern, his political sympathies increasingly carried him away from his former mentor. To the extent that Tarnier had expressed interest in the depopulation question at all, he had usually done so in the conservative sense of offering incentives for mothers producing large families. During a visit to the small community of Arc-sur-Tille, for instance, he publicly offered one hundred francs to every family bearing a new infant within the following year; his pleas were not in vain, for eighteen families (out of eight hundred inhabitants) responded.[20] Budin's political sympathies gravitated to the political left-of-center, manifested by speeches to socialist groups in the 1890s followed later by close associations with solidarist leaders such as Paul Strauss.[21]

Far more than had Tarnier, Budin tapped into his country's growing frustration over depopulation as he developed his own professional career. Even before his arrival at the Maternité in 1895, he had already received considerable attention for developing the *consultations des nourrissons,* a system of clinics supervising the care of newborn infants discharged from maternity hospitals. He launched the first *consultation* while obstetric chief of the Charité hospital in 1892, where he feared that many newborns left the hospital only to die in the crowded dwellings of Paris. The idea proved remarkably successful; by 1905, some twenty-eight similar institutions would operate in the French capital alone. Like most professional contemporaries, Budin saw bottle-feeding and resulting infantile diarrhea as the main contributors to this mortality; where he stood apart (especially compared to Americans) was his insistence on the superiority of breast milk to even the best substitutes. The primary purpose of the *consultations* was to encourage mothers to continue to nurse their infants after discharge by providing advice and monitoring their weight.[22] In some cases, *consultations* provided economic incentives, such as gifts of free meat or a gratuity rewarding faithful attendance.[23] Budin saw his career as the continuation of Tarnier's program; he advocated the expansion of the obstetrician's sphere beyond the mother to the newborn and indeed to the older infant prior to weaning. The extension of French obstetrics to the realm of the nursling imparted an orientation to physicians such as Budin distinctly different from that of either American obstetricians or pediatricians.[24]

Budin took equal advantage of the greatly enhanced status of medical science in the wake of the germ theory as he sought professional authority outside of the hospital sphere.[25] In France during the late nineteenth century, the name of Louis Pasteur symbolized the revolution overtaking medical practice. Pasteur's flair for spectacle, whether through the public demonstration of the anthrax vaccine at Pouilly-le-Fort in 1881 or the vaccination of the young rabies victim Joseph Meister in 1885, made the promise of medical science tangible as never before.[26] Scientific positivism, moreover, was the order of the day in the context of another drama, the struggle between the Third Republic and the Catholic Church. The expulsion of religious orders from the Paris hospitals, noted with disapproval by American observers, provided a particularly vivid example of the ideological climate pervading French medicine.[27] The hygienic reformer Paul Strauss, citing Budin's collaboration with Tarnier in the "Pasteurian victory" over puerperal fever, praised him as an altruist who constantly sought the goal of "placing science in the service of humanity."[28] Another associate physician honored him for never advancing anything without "proof."[29] Budin's identification with positivism and the legacy of Pasteur proved to be great assets in his becoming a national advocate for the premature infant.

Pierre Budin thus brought a set of values and experiences to the Maternité that went beyond those of Tarnier. Both shared the common experience of having participated in the near eradication of puerperal fever, and both were equally ardent devotees of Pasteur and the germ theory. Budin nonetheless went further than his mentor in showing an even greater predilection to encompass both the infant and the mother in his reform efforts. In the process, he sought to engage the contemporary social concern over depopulation, to view his mission in a framework going beyond the confines of the hospital. Budin's background made him a key figure in integrating the incubator into the broader French social context. His serious involvement in this regard began with his experience in the Maternité.

The Pavilion of Weaklings

Budin took over the Maternité in 1895 and remained in charge for three years until his promotion to full professor in the Paris Faculty of Medicine in 1898.

TABLE 3.1
Mortality Statistics, 1893–1903

Year	Patients Treated	Place of Birth[a] Inborn	Place of Birth[a] Outborn	Deaths (% Mortality)
1893	211	57	154	87 (41%)
1894	563	121	398	252 (45%)
1895	445	69	226	258 (58%)
1896	377	54	311	241 (61%)
1897	391			292 (75%)
1898	482			343 (71%)
1899	492			322 (65%)
1900	730			436 (60%)
1901	687			502 (73%)
1902	757			587 (78%)
1903	568			400 (77%)

Source: Data from "Registre des Enfants Débiles, 1895–1903," series 6Q2.1–2. Assistance publique, Service de la documentation et de archives, Paris.
[a] Unavailable after 1896.

His relatively brief appointment nonetheless left an indelible mark on his later attitude toward the hospital care of premature infants. Budin arrived just as the high expectations surrounding the *pavillon des débiles* were crashing against the reality of soaring mortality rates. Tarnier in the 1880s had claimed to reduce premature infant mortality from 66 to 38 percent, and in fact the overall mortality rate of the new service in 1893, 41 percent, was not far above that figure (see table 3.1).[30] But during the next four years, it rose to 74 percent and remained in this range for most of the five years thereafter while records were still maintained. To some extent, the institution was a victim of its own popularity. Annual admissions quickly doubled and indeed tripled by 1899. As expected, babies admitted from outside the hospital accounted for the great majority of the increase in admissions during the first four years. The heavy patient load presented obvious challenges for the hospital personnel; Budin noted in his journal with regard to two visitors in July 1896, "Both are astonished upon seeing the *service des débiles*. . . . They also see that this service is no sinecure."[31] Yet the problem was not simply one of an overwhelmed staff, for Budin as well as other observers testified that the nurses maintained standards of asepsis and care second to none.[32]

On one level, the reason for the increased mortality was fairly simple: the institution received an extremely ill population of patients. Infants repeatedly arrived at the Maternité in such a debilitated state that their fate was a foregone conclusion. Budin was especially struck by the grim outcome of those infants who were chilled prior to arrival. Even when treated with hot water baths followed by incubator therapy, these infants rarely recovered. Infants whose temperature on admission was less than 32°C (89.6°F) suffered a 98 percent mortality, while those with temperatures between 32°C and 33.5°C (89.6°–92.3°F) fared only slightly better, with a 90 percent mortality. Nearly a third of Budin's patients fell into one of these two categories.[33] In addition to the effects of exposure, roughly two-thirds of the infants manifested clinical symptoms beyond simple debility, including many of the diseases of prematurity (such as sclerema, syphilis, cyanosis, and jaundice) discussed earlier as well as infectious conditions, including enteritis and diarrhea.[34] Budin responded to this situation by employing what became a common practice: dropping from statistical analyses those infants who had died within forty-eight hours of admission or who had rectal temperatures of less than 32°C.[35] He clearly believed that many of these infants were beyond hope of intervention. The incubator was a preventive therapy, not a cure for the infant who had already been chilled.

More controversial was Budin's interpretation of why so many infants arrived severely ill. He did not envision the problem as simply a matter of cold exposure sustained by the trip to the hospital, a problem of premature infant physiology and the logistics of transportation. Indeed, it was far from clear whether parents were trying to rush their babies to the Maternité in the first place. Most infants were admitted after a period of days rather than hours.[36] In commenting on the high percentage of deaths in the first forty-eight hours after admission, Budin wrote that "too often the *service des débiles* served only as a mortuary depot. The more the service became known, the more it seemed to constitute, for many people in the city, a place where one transported his little infant when it was going to succumb."[37] Confirming this impression was what Budin believed to be the carelessness shown by parents in transporting their infants to the hospital. He described one example of a premature infant transported over a hundred miles with no more covering than a piece of flimsy material as bordering on "infanticide by neglect."[38]

Budin's "evidence" that infants brought to the hospital were victims of neglect was, of course, highly open to interpretation. Parents may well have been unaware of the critical importance of preserving the baby's temperature. Unfortunately, we are for the most part left to speculate on their motives. The publications and surviving archives relevant to the *pavillon des débiles* offer no direct testimony from mothers. But testimony does exist from the physician who later replaced Budin as head of the service in 1898 and supervised it until its closure in 1903, Charles Porak.[39] During his years in charge of the service, its volume and overall mortality rate rose still higher. Like Budin, Porak found that the great majority of the premature infants brought from home were already severely compromised; they usually arrived two or three days after birth but sometimes after as many as fifteen days. Yet Porak interpreted this delay as evidence of maternal attachment. "It is only after having vainly tried to save them," he wrote in 1902, "after all hope is definitely lost, that the families decide to separate themselves from them."[40] Whether Budin or Porak was closer to understanding the mothers of their hospitalized infants is difficult to say; neither probably had much direct interaction with the mothers themselves. The arguments of both, however, echo a broader dispute within French society over the issue of infant abandonment.

One of the features that marked the depopulation debate in France was the extent to which it centered on the responsibility of the mother for the nation's high infant mortality rate. This emphasis reflected the country's efforts to deal with its legacy of wet-nursing and foundling hospitals. The practice of sending infants to the country to be nursed was far more widespread in France than any other western European country and remained commonplace among the working classes until the First World War. Reformers believed it to represent a particularly unnecessary contributor to infant death, and indeed the first infant welfare legislation in France, the Roussel law of 1874, provided for its regulation.[41] The continuing existence of another holdover from the eighteenth century, the foundling hospital, provoked equally contentious debate. Paris's Hôpital des Enfants Trouvés remained a focus of controversy throughout the 1800s, as moralists and reformers debated whether the institution's existence encouraged abandonment or prevented infanticide.[42] In parallel with the increasing urgency of the depopulation problem, the latter part of the century witnessed a trend replacing moral exhortations with tangible programs to counteract abandonment by educating and supporting mothers to care for their infants. Reformers spoke of abandonment pragmatically as a product of an industrial society forcing women to work outside the home. In the new climate, the Paris foundling hospital was renovated and transformed into a pediatric hospital and renamed the Hospice des Enfants Assistés.[43] But the older moralistic language rarely disappeared from the language of even more sympathetic reformers such as Budin, who continued to frame infant mortality in terms of the promotion of motherhood.

There were, in fact, some striking parallels between the Maternité's service for weaklings and the older foundling hospital that underlay Budin's concerns. Like a foundling hospital, the service received infants whose birth weight or length of gestation was unknown. Infants had to be classified according to admission weight, which might reflect malnutrition rather than prematurity. An eighteen-hundred-gram infant at admission might have been born at twenty-five hundred grams.[44] It made sense to call the institution a service for weaklings rather than for premature infants. This classification also suggested a connection with the legacy of the foundling hospital, whose physicians had long blamed their patients' high mortality rates to some extent on the high proportion of *débiles* that filled their wards.

Indeed, premature infants may have accounted for much of the mortality at Enfants Trouvés. Although the institution in theory operated as only a temporary way station for abandoned infants awaiting placement with rural wet nurses, many infants never survived to be discharged. During the first half of the nineteenth century, roughly a quarter of all infants admitted to Enfants Trouvés died during their short three- to four-day stay. The hospital environment certainly played a part in this mortality; wet nurses were constantly in short supply, infants were often crowded two to a crib, and the poorly heated wards were plagued by drafts and temperatures that dropped below 50°F in the winter.[45] Contemporary physicians believed, however, that many of the abandoned infants were *débiles* who were already compromised upon admission. An investigative commission in 1860 attributed most of the deaths at Enfants Assistés to weakness upon arrival.[46] Though the term *débile* did not become synonymous with prematurity until late in the century, it is interesting that by Budin's time a historical account of foundlings in a popular medical journal asserted explicitly that premature and other *débile* infants accounted for almost half of all the admissions of abandoned infants in the early 1800s.[47] The point to emphasize is that many foundlings were also classified as weaklings. Enfants Trouvés conceivably provided parents faced with a dying newborn with a socially sanctioned means of not having to witness that death first hand. From Budin's perspective, the *service des débiles* at the Maternité inherited a similar function from the standpoint of many of the city's inhabitants. In trying to create a modern hospital nursery, the hospital had instead reincarnated the foundling home.

Given these reservations, it is not surprising that Budin's own critique of the service centered on the symbol of degenerate motherhood, the wet nurse. Although the *service des débiles* incorporated four professional nurses, it remained heavily dependent on roughly three times as many hired *nourrices* for feeding and for much practical care.[48] At best, Budin harbored mixed feelings, if not outright paternalism, toward these women. He displayed great interest in documenting their milk production, showing, for example, that it was possible for one wet nurse to provide thirty-four feedings a day to hospitalized premature infants in addition to her own child. Budin also shared his generation's belief that the wet nurse's diet and emotions could influence the nursling and therefore insisted upon keeping the *nourrices* in the hospital, where their eating habits and morals could be strictly supervised. However typical for someone of Budin's background, these various restrictions clearly had an oppressive side that provoked resentment and antagonism.[49]

The wet nurse came to represent for Budin a persistent source of irritation and a convenient scapegoat for the institution's problems. The *nourrices* actually went on strike in late 1895, when Budin tried to prevent parents from luring them away from the hospital when their infants were discharged.[50] The Maternité survived this crisis by temporarily relying on cow's milk; more serious in the long run was the tendency of wet nurses and their infants to introduce infection into the ward. In spite of all prohibitions, some resourceful *nourrices* and their infants managed to evade their confinement, often acquiring a respiratory illness in the process. Although these infections typically caused only minor symptoms among the nurses' infants, their impact on the premature babies could be devastating. Budin linked no less than four epidemics of bronchitis between 1895 and 1897 to the wet nurses. The mortality from this source peaked in the last of these three years, when two epidemics, each lasting a month, killed a total of thirty-four infants. It should be noted that these epidemics accounted for only a small fraction of the service's total mortality, and at any rate the premature infants themselves represented a source of infection as well. The problem nevertheless spurred Budin to redesign the service in order to house the wet nurses' living quarters in a separate pavilion and to allow individual isolation of all suspect as well as contagious infants.[51] The impact of these reforms is difficult to assess, since they were not completed until Budin had left the hospital in 1898. Their benefit may have been countered by the willingness of the new management to accept still more infants, since hospital mortality continued to rise in spite of the reduction of epidemic infection.[52]

What ultimately came to disturb Budin most about the practice of wet nursing was its interference with the relationship between the mother and her infant. The mother's reluctance to give up an infant to the hospital was understandable, given her complete loss of control to the nurses and wet nurses for its care. "I have been grieved to see a certain number of women come more and more rarely to visit their child," he observed, "and gradually lose all interest in it, abandoning the child."[53] Even where outright abandonment had not occurred, infants faced a questionable future when discharged to their parents. The infant left its *nourrice* for a mother no longer capable of nursing it; as a consequence, it had to be fed by bottle or sent to another wet nurse. In 1897, Budin took the remarkable step of sending letters to his patients' mothers to assess the fate of all the premature infants discharged from his service during the earlier part of the previous year. His survey revealed a mortality rate ranging from 15 percent among the infants with wet nurses to 41 percent among those fed by bottle.[54] The hospital's work, it seemed, could be nullified not only by the infant's condition at admission but also by its treatment at home after discharge.

Though the Maternité's *service des débiles* may in retrospect appear to be the first special-care nursery, it eventually struck Budin as a misguided enterprise. He thus turned the institution over to his successor Charles Porak in 1898, who continued to defend the service on the grounds that it fulfilled a need for women unable either to care for their sick infant or to enter the hospital themselves. Porak pointed out, quite fairly, that his poor mortality statistics could not be compared with those of maternity hospitals, since the service had been transformed from one for healthy infants to one for sick premature infants.[55] Yet he never became an important advocate for premature infant care and appears to have lost interest after the service was closed in 1903. The main direction of reform by this point was moving beyond the hospital care of premature infants. The question of long-term outcome now acquired central importance, as obstetricians scrambled to develop alternatives to the ill-fated service for weaklings.

From Technology to the Mother

The setbacks at the Maternité led to a broader challenge to premature infant care throughout the Paris medical world. Interestingly, the incubator itself played only a secondary role in this controversy. The device's detractors were limited, for the most part, to France's small pediatric community. In 1898, the pediatric specialist Antonin Marfan asserted that it was not possible for other physicians to share obstetricians' enthusiasm for the *couveuse.* Although he repeated earlier concerns over the problem of maintaining a stable temperature, he concentrated his attack on the problem of infection. The employment of a thick cotton filter to exclude infection, which Marfan believed necessary, compromised the ability of the incubator to circulate air by convection. Yet without such a filter, he asserted, "the apparatus gathers and accumulates all the poisons of the atmosphere around the infant; this is without doubt the source of the bronchopneumonia which so frequently strikes the prematures."[56] Other pediatricians agreed that the incubator, though not in itself a source of infection, could nonetheless increase the risk of infection in an unsanitary environment.[57]

One response to the ventilation critique was the development of a new generation of incubators that addressed these concerns by actually drawing their air supply from outside the hospital. The most widespread of these was the Lion incubator, developed in 1891 in Nice, France. The Lion was an elaborate metal incubator designed to be ventilated with outside air forced through a system of intake pipes. This device would become an important alternative to the Tarnier *couveuse* both in Europe and the United States.[58] But Budin specifically condemned the Lion as complicated and unreliable.[59] Indeed, it evolved along a pathway largely independent of the Paris maternity hospital. We will return to the Lion when we consider the transfer of incubator technology from France to the United States. For the moment, the main point to underline is how little it influenced the Paris obstetric leaders who dominated the French discourse over the incubator.

The ventilation critique, at any rate, never offered as great a challenge to the Tarnier incubator in France as it would in the United States. Many Paris obstetricians in the wake of asepsis went to great pains to distance themselves from the atmospheric theory of infection implicit in Marfan's assertions. Their strong identification with Pasteur no doubt reflected the way in which their own specialty had been transformed through the reduction of puerperal fever.[60] But even the response of most pediatricians was generally more nuanced than that of Marfan. The director of Enfants Assistés, Victor Hutinel, created a *pavillon des prématurés* in the 1890s by evacuating and disinfecting the hospital's older nursery for syphilitic infants.[61] The resulting institution represented a kind of compromise that retained a place for the incubator in young premature infants while placing older ones in open cradles. His nursery was furnished with incubators and isolation facilities similar to those of the Maternité but surrounded by gardens where older premature infants could be exposed to open air and sunshine. Hutinel fittingly called his outdoor treatment "the air cure, the sanitarium of newborns," and indeed it seemed to assume a fundamental similarity between the weakness of prematurity and that of tuberculosis. He nonetheless recognized a legitimate role for the incubator in the younger and smaller infants. The device could still be used advantageously if suitable precautions against infection were taken. "Tant vaut le milieu, tant vaut la couveuse," he wrote in 1899: the incubator was as safe as its environment.[62]

Another factor that may have prevented the ventilation critique from becoming more widespread in France was that pediatricians, who tended to view it more sympathetically, played a far smaller role in developing the incubator than they would in the United States. No contemporary French pediatrician began to approach the dedication of Tarnier or Budin in promoting premature infant care. Their relative lack of involvement in early infancy mirrored the stronger claims by obstetricians such as Budin to the supervision of the infant prior to weaning.[63] The response of French pediatricians is interesting for its parallels with the American pediatric profession, a theme to which we shall return. But it was, in general, the obstetric profession that dominated the terms of the debate over the premature infant in France. While French pediatricians might enumerate the same list of objections to the incubator as would Americans, in the end they generally conceded the device to be efficacious so long as it was used with proper precautions.[64]

The focus of controversy in France was not so much the incubator as the question of whether premature infant mortality could best be counteracted through treatment or prevention. Reformers moved in two directions, both outside the hospital; one party encouraged long-term support, the other prenatal care. The leading advocate of the latter approach was Adolphe Pinard, Budin's greatest rival and professor of obstetrics on the Paris Faculty of Medicine. Pinard directed the new Baudelocque maternity hospital adjoining the Maternité, where he would have been well aware of the disappointing results of the *pavillon des débiles.* Along with Budin and the pediatrician Gaston Variot, he was one of the most influential physicians involved in the French infant mortality campaign at the turn of the century.[65] Again, the prominence of Budin and Pinard reflected the involvement of French obstetricians in supervising the infant as late as the second year of life. Pinard was known as an advocate of maternal support programs in reducing infant mortality; he promoted prenatal care as well as the teaching of *puériculture* (the science of child rearing) to young women. He also became a founding member (and later president) of the French Eugenics Society. The context of depopulation anxiety helped to impart a distinctive tone to eugenics in France, one which emphasized positive measures to promote the health of future generations over negative efforts to restrict the marriage and propagation of the "unfit."

Pinard's sympathy for the neo-Lamarckian doctrine of the inheritance of acquired traits led him to see the supervision of the mother during pregnancy as potentially having great influence on the development of the fetus.[66] In this respect, he concentrated on countering the adverse effects of urban employment on fetal development. In 1895, he reported to the Academy of Medicine a study comparing the birth weights and gestation periods of one thousand infants born to mothers who worked until term with a population of equal size whose mothers had spent the latter part of pregnancy in a municipal shelter. Nearly twice as many of the women who worked until delivery (236 versus 126) gave birth to infants prior to thirty-eight weeks' gestation. These dramatic results, though uncontrolled for other differences between the two populations, prompted Pinard to conclude that "if we want a strong and vigorous population," the French nation would need to move beyond treating the feeble and sick and prevent premature birth by providing for maternity leave.[67]

In the wake of Pinard's study, Budin faced the challenge of redefining the problem of postnatal care for the premature infant all the more acutely. He had his chance to develop an alternative on Tarnier's death in 1897, when he inherited his former teacher's position as professor of obstetrics at the Faculty of Medicine. The new appointment moved Budin from the Maternité to a newer maternity hospital, designated the Clinique Tarnier in honor of his predecessor. There he rejected the idea of creating a service for premature infants, an enterprise that he had found required great time and ultimately yielded only meager results.[68] Budin nonetheless still had to contend with the substantial number of premature babies born within the institution, a population representing about 10 percent of all births. By applying the same principles he had used at the Maternité, Budin was able to obtain, apparently to his surprise, a much higher survival rate. As a result, he began to formulate a substantially different rationale for premature infant care that carried him beyond the tradition of Tarnier. Although Budin continued to emphasize incubators, feeding, and infection control (all issues addressed at the Maternité without success), he now called attention to an additional factor he believed at least as crucial.

The essence of Budin's approach at the Clinique Tarnier was the integration of the mother into her infant's care. He brought the incubator and the mother together into what might be called a mother-infant nursery. The Clinique Tarnier, like the *consultations*, became known as a "school for mothers," an institution that extended its sights beyond preserving the infant's course in the hospital.[69] Infants were kept in glass incubators at their mothers' bedside on the ward; the mother literally functioned as nurse. In keeping with Budin's concerns regarding the long-term disadvantages of wet nursing, the hospital followed the example of the *consultations* in requiring maternal breast-feeding. Since this policy presented obvious difficulties for infants too weak to demand milk on their own, Budin took the pragmatic if unsentimental strategy of having a mother and wet nurse temporarily switch infants until the mother's milk supply came in. Many premature infants too weak to stimulate the initial production of breast milk nonetheless could still take milk from a mother whose supply was well established; others who required gavage could eventually be returned to a nursing mother as well. Budin thus managed to ensure breast-feeding in all but a few of the hospital's premature infants.[70]

Underlying all these reforms was the conviction that rescuing the infant was meaningless without the active involvement of its mother. Contemporaries recognized another advantage of Budin's system to be the encouragement of maternal attachment and devotion to an infant who might ordinarily have been abandoned. In the rather melodramatic language of one journalist who had seen the hospital's incubator service,

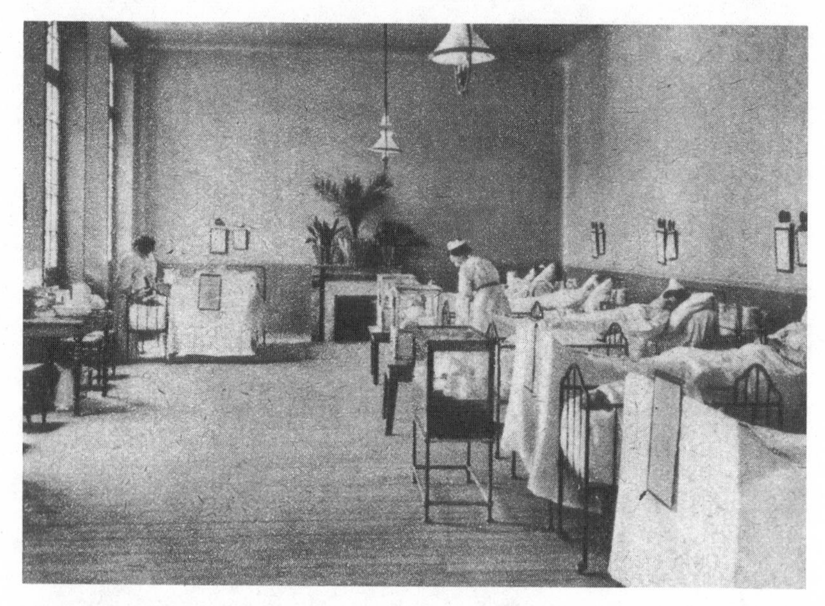

Mothers and infants in Paris maternity ward, c. 1900. *Source:* Administration generale de l'assistance publique, *Cent ans d'assistance publique à Paris* (Paris: Administration générale, 1949), 21.

> The glass cover permits the mother to watch every movement of the poor, fragile little being. And thus by watching him, almost minute by minute, the mother becomes attached to her baby; she trembles for him during the weeks she remains at the Clinique . . . [and after discharge] she will return him with faithfulness to the *consultation de nourissons.*[71]

The point of the mother-infant ward was to improve long-term outcome. Budin continued to follow his infants after discharge through a *consultation* established at the Clinique Tarnier. He stated his philosophy succinctly in his 1900 text, *Le Nourrisson:* "First, save the infant, the essential point; second, save it in such a way that when it leaves the hospital it does so with a mother able to suckle it."[72]

The divergent approaches of Budin and Pinard finally collided when both testified on separate occasions before the Senate Commission of Depopulation in 1902. Pinard seized the opportunity to bring the dismal record of the Maternité's *service des débiles* before the public. Though he was not the first to express reservations publicly, his testimony in this forum represented the most visible such challenge to premature infant care mounted in France thus far.[73] Pinard began by citing the climbing mortality statistics of the institution from

41 percent in 1893 to 73 percent in 1901. He then asked, somewhat rhetorically, what fate was in store for the survivors. "With only very rare exceptions," he pronounced, "these prematures, these weaklings, for whom so many sacrifices have been made, who have cost so much care . . . remain for the durations of their lives weak or infirm."[74] The implication was clear: the resources of government and philanthropy were better spent on prevention than on treatment.

Budin drew on the experience of the Clinique Tarnier to counter Pinard's charges. He presented a series of remarkably low annual premature infant mortality rates from the institution that ranging from 4.8 percent to 17.8 percent.[75] Admitting that results this favorable had initially struck him as almost illusory, Budin affirmed their validity by citing a similar experience (11.5 percent mortality among 368 premature infants) reported by his colleague Charles Maygrier at the maternity department of the Paris Charité hospital.[76]

In response to Pinard's dismal prognosis for those premature infants who survived the hospital, Budin offered the results of his own efforts at follow-up. Over the previous several years, he had gone to great lengths to keep up with his infants after discharge. The task was complicated, however, by the fact that only a minority of mothers (presumably those most motivated to breast-feed their infants) remained in the *consultations.* Budin and his students attempted to trace many of the others but often lost track of those whose addresses had changed. Because public records enabled them to find most of the infant deaths, the overall long-term mortality rates for those infants who could be located were almost certainly inflated. As of 1901, for example, Budin reported that he had been unable to find 65 percent of the premature graduates from the Clinique Tarnier in 1898; of those he could locate, a third had died.[77] In his writings Budin countered such figures to some extent with anecdotal stories of prodigies, such as a seven-year-old graduate who could now read French and German.[78] But the principal refutation was provided by the record of the *consultations,* which he believed set the standard for what could be achieved. In his report to the commission in 1902, he reported that only one of the sixty-six premature infants followed had died.[79]

It is worth noting that although both sides of the controversy put great stock in statistical arguments, neither had really found any objective answers to the many questions regarding the long-term outcome of the premature infant. The problem was far more complicated than suggested by either Pinard or Budin. The two obstetricians were comparing fundamentally different populations. Nearly 80 percent of the Clinique Tarnier incubator patients weighed more than two thousand grams (see table 3.2). Tarnier and Auvard, it will be recalled, had justified the incubator's efficacy primarily with regard to premature infants with weights below that same threshold. The preponderance of larger patients probably helps explain why Budin

TABLE 3.2
Premature Infant Survival Statistics under Pierre Budin, in Paris Maternité, Service des Débiles (1895–1897) and Clinique Tarnier (1898–1900)

Weight (grams)	Number of Patients (% Survival)	
	Maternité	Clinique Tarnier
<1500	186 (12.9%)	13 (69%)
1500–2000	432 (38.6%)	78 (76.9%)
2000–2500	212 (67.5%)	344 (91.2%)
Total	830 (40%)	435 (89.6%)

Source: Data from Pierre Budin, *The Nursling: The Feeding and Hygiene of Premature and Full-Term Infants,* trans. William J. Maloney (London: Caxton Publishing, 1907), 63; Pierre Budin and M. Perret, "Nouvelle recherches sur les enfants débiles," *Obstétrique* (1901): 210.
Note: Figures exclude deaths in first forty-eight hours.

recommended lower incubator temperatures than did virtually all of his contemporaries—typically in the 77–78°F range, well under Tarnier's recommendation of 86°F or higher.[80] Even within a given weight category, the two populations were still far from comparable. The Maternité's figures represented weights at admission, not birth weights. Its infants suffered from varying combinations of prematurity, exposure, illness, and malnutrition.

In the final analysis, Budin had not so much proved the worth of the incubator as defined a target population in such a way as to portray his interventions as constituting a medical breakthrough. Perhaps the most graphic example of this process was his continuing practice of excluding from his statistics babies who had died within the first forty-eight hours of life.[81] He argued that this was necessary to compare his population with that of the Maternité; but the result was to produce overall survival rates so high as to make Budin's interventions appear impressive in absolute as well as comparative terms. Budin never tested whether a padded basket with hot water bottles would work as well as a simple incubator for his restricted population of relatively large premature infants. Although his survival rates were hardly better than those of some contemporary Americans in maternity hospitals if adjusted for patient population, Budin achieved the status of an international authority on the care of premature infants.[82] He understood the power of dramatic statistics to influence the medical profession and laity and used them to campaign vigorously for the premature infant.

The last phase of Budin's career witnessed his increasing influence on the French infant mortality campaigns at large, helping to endow them with a neonatal emphasis distinct from that in the United States. Whereas contemporary American infant mortality reformers focused almost exclusively on the prevention of diarrheal disease in the early 1900s, Budin sought to extend the French discourse to prematurity as well. For example, the organization of his textbook on infant feeding, *Le Nourrisson,* published in 1900 and translated into English in 1907, differed dramatically from its Anglo-American counterparts in devoting nearly a third of its content to the special problems of premature infants.[83]

Budin brought a similar emphasis on the problems of the newborn to the Ligue Contre la Mortalité Infantile, which he founded in 1902 along with his political ally, Senator Paul Strauss. He thereafter became involved in a series of infant mortality surveys in which he consistently highlighted the importance of addressing congenital debility as well as gastroenteritis. For example, his examination of infant mortality in Paris revealed that this problem was the second leading cause of death in infancy, accounting for 170 deaths per thousand, as opposed to 384 caused by gastroenteritis. In an address to the league in 1903, he asserted that although it had once been believed that such congenitally weak infants were not worth the trouble to raise them, "Scientifically, we must today affirm the contrary. It is possible, in great proportions, to save these infants."[84] Budin joined Strauss in promoting a bill in 1902 to provide a variety of infant welfare measures including government sponsorship of maternity leave, pregnancy rest homes, and infant *consultations.* Part of its provisions specifically compensated families for the care of premature infants, including the use of incubators, in the home or hospital.[85] Though such measures remained fairly modest, and comprehensive maternity leave legislation awaited 1913 before passage, they affirmed an important principle that would increasingly distinguish France from the United States: economic assistance was necessary, as well as education, to support mothers in caring for their infants.[86]

Budin's career came to an abrupt end in 1907. On January 13, he gave an address on infant mortality in Marseilles, a speech in which he elaborated on the frequency of congenital debility in the provinces and urged its treatment. While at the conference, he was incapacitated by influenza. On January 22, Budin died at the age of sixty-one.[87] His former rival Adolphe Pinard became the leading obstetrician involved in the French infant mortality movement. Pinard's rise to some extent shifted emphasis away from the treatment of prematurity back to its prevention. The transition was less dramatic than might have been expected, however, in part because the differences in the approaches of the two men had long masked two underlying similarities. Both were concerned over the long-term outcome of the premature infant, an outlook in marked contrast both with that of Tarnier twenty years earlier and with most American incubator advocates of the time. They also shared a conviction that any attempt to reduce premature infant mortality that would not be reversed after leaving the hospital had to include the mother. Their disagreement was not over the centrality of the mother's influence on the health of her infant but over whether public health efforts should be concentrated before or after birth.

Budin's emphasis on inexpensive technology in neonatal care and his concentration on relatively well-developed premature infants removed much of the basis for the conflict. In the end, Budin advocated maternity leave just as did Pinard, and the foundation established in his name incorporated both the *consultations* and Pinard's idea of *puériculture* classes.[88] His philosophy was carried on by his former student Charles Maygrier, who had become chief obstetrician in charge of the Maternité by the time of Budin's death. A devoted admirer of Budin, Maygrier conducted his own studies of the long-term outcome of premature infants to counter charges that they were "social waste products."[89] Maygrier's work was largely derivative, however; his contribution was to preserve a niche for the incubator in an obstetric world increasingly drawn to Pinard's emphasis on prevention. Pinard, for his part, allowed the *couveuse* at least a peripheral role in his infant mortality strategy. In a 1908 tribute to Tarnier ten years after the latter's death, Pinard allowed that though the old master had overestimated the potential of the incubator, the device nonetheless often provided premature infants a "safeguard against sickness or death."[90]

Nonetheless, although the medical leadership of France acknowledged the role of the incubator in the treatment of prematurity, it showed little interest in its further development. The prevention of prematurity attracted far more attention. Although its story cannot be told here, a few words about Pinard's later career are in order. At the onset of the First World War, Pinard became president of France's newly established Central Office for the Assistance of Mothers, where he oversaw the expansion of maternity homes and *consultations* that provided incubators. Yet he remained best known as an advocate of prenatal care and maternity leave. In 1916, he criticized the employment of women in factories during the war, complaining that "it takes longer to produce babies than to make shells or cannon."[91] By then the cataclysm of war was taking its toll on the accomplishments of both men. The infants of the *service des débiles* from the 1890s had now reached the age of soldiers; the nation that had originally saved them now returned to reap its investment. The battalion of infants for which Budin was eulogized as preserving now faced, as one historian has ironically noted, the trenches of the

Western Front.[92] The years of French leadership in neonatal medicine had come to an end as well.

❧

Budin's landmark text *The Nursling* became available in English in 1907, the year of his death. The book provided the most thorough account of premature infant care available in its time, synthesizing much of the work of Tarnier, Budin, and the French obstetric tradition. Indeed, along with the Clinique Tarnier statistics it eventually helped to win Budin a reputation as one of the fathers of neonatal medicine.[93] Budin's relationship to the later evolution of newborn care was in fact more complicated. He was unquestionably the most vocal advocate of premature infant care at the dawn of the twentieth century and obtained survival rates exceeded by none. Yet his specific innovations are difficult to fit into any chronology of technological progress. Although Budin had experimented with thermostatic incubators in the 1880s, by the turn of the century he rejected the best such model available (the Lion incubator) in favor of the simpler Tarnier-Auvard *couveuse.* Indeed, he undermined his predecessor's rationale for the incubator by focusing on larger infants less vulnerable to heat loss. Budin even continued to employ the labor-intensive expedient of hot water bottles to heat the device in place of a heated-water reservoir.

Budin's contributions are best considered not in terms of a step forward or backward but as one branching off in a different direction. Instead of seeing the incubator as potentially replacing the functions of mother and nurse, he saw it as extending them. His most characteristic attitude toward incubator design was symbolized by the construction of the incubator chamber almost entirely of glass, making its occupant visible to mother and physician. The apparatus provided a means of allowing the infant to coexist with its mother independent of the temperature of the surrounding ward. It was, in other words, a bedside technology. The device's requirement for close attention reinforced the mother's responsibility for the child. Budin's accomplishment was to look beyond the confines of both incubator and hospital, promoting the device as part of a larger program providing long-term support to the mother. His interest in maternal involvement and long-term care over technological intervention might be labeled as a supportive style of medicine distinct from prevention and acute care. This style reflected Budin's professional values, derived from his composite background as obstetrician and infant welfare advocate. It also echoed the context provided by France's discourse over depopulation, an orientation that envisioned the promotion of motherhood as a patriotic duty.

CHAPTER FOUR

Technology Transfer and Transformation

In the course of moving from the context of Paris maternity hospitals to the United States, a remarkable transformation overtook the incubator. The humble *couveuse* became a complex and automatic machine. It would not be much of an exaggeration, in fact, to say that during the late 1800s the device evolved in opposite directions inside and outside the French obstetric tradition. The main thrust of Tarnier's and especially Budin's innovations had been to simplify the incubator, to integrate it into its institutional and cultural context. Whether or not the device even deserved to be labeled as a technology could be argued; certainly, it was no more than a peripheral technology. The new generation of incubators that emerged during the 1890s, on the other hand, belonged unambiguously to the domain of machines. With the aid of such mechanical adornments as thermostats and ventilation systems, they provided a far more autonomous technological environment than had their predecessors. One admirer proclaimed that "the day for boxes and baskets heated by hot bottles, hot sand, and what not, serving as incubators, has gone by."[1] The refined incubator provided a scientific means of caring for the premature infant.

One is tempted to recount the evolution of these complex incubators in terms of technological progress. They did, in fact, incorporate many ingenious features that seemed to place them ahead of their time. But progress is often more easily discerned in hindsight. Far from advancing along a line toward our own "modern" conception of the incubator, the device branched out in a variety of directions. The French obstetric tradition represented merely one such possible pathway, one whose members condemned many of the features of the new generation of incubators. Indeed, the innovators whose models most resembled those of today were often furthest in spirit from the professional mainstream of their own countries. Complex incubators lost favor around the time of the First World War nearly as rapidly as they had ascended twenty years earlier. As seen from the perspective of contemporaries, they came to represent not so much the birth of neonatology as a therapeutic fad.

The incubator thus evolved in a branching pattern as it left its Parisian origins. At this point it entered the most shadowy phase of its evolution, a phase in which it had left the Paris maternity hospital context but had not yet been incorporated into a new professional context in the United States. So few were the direct connections between the medical leadership in the two countries that third parties played a critical role as intermediaries. Among such agents of transfer were professional inventors, instrument suppliers, and entrepreneurial physician-inventors. Their involvement helps explain why the peak of American interest in the incubator occurred in the early 1900s rather than at the time of Tarnier's first papers twenty years earlier.

The importance of the contribution of this technological tradition of incubator design can be highlighted at two levels. First and most obviously, American obstetricians and pediatricians by the late 1890s for the most part were no longer responding directly to the Tarnier *couveuse* but to its transformed successors. At the same time, the metamorphosis of the incubator involved more than mere hardware. On a second level, its reconstruction engaged social and cultural as well as mechanical factors. The device acquired new associations and meanings linking it to technology and the various connotations of that word.

Inventors, Imagination, and the Incubator

Medical journals were in all probability the most important means of introducing incubator technology to the United States during the first ten years after its invention. Leading British and American journals frequently maintained correspondents in Paris and Berlin and printed short abstracts of important contributions from those countries. A wave of such articles appeared between 1883 and 1886 following the first published reports by Stéphane Tarnier, Alfred Auvard, and Carl Credé. Virtually all cited the critical statistic in Auvard's paper, the drop in premature infant mortality from 65 percent to 38 percent with the incubator. On the matter of constructing the device, however, these abstracts varied considerably. Some were quite detailed; the *American Journal of Obstetrics and Diseases of Women and Children* reproduced Auvard's original diagrams with a detailed description.[2] Most writers were content to provide a sketchy account of the simplified Tarnier-Auvard design. These reports were important, because they appear to have been the only direct source for Americans seeking to build their own versions of the device throughout the decade. Although by 1884 British obstetricians could import Auvard incubators directly from France, Americans had little if any such access until the 1890s.[3] The *couveuse* was nonetheless so simple that published descriptions went far to make possible its adoption. As early as 1886, a reviewer in the *American Journal of the Medical Sciences* asserted that the device was "too well known to need a description."[4]

Yet building an incubator was more straightforward than understanding its operation. In 1884, the *Medical Record* stated that during the two years since Tarnier had introduced his invention, "various and incomplete descriptions of this apparatus have been published, and strange misconceptions of what the *couveuse* is, and what it is intended to do, have got abroad." The writer went on to assert that the device was not for the cultivation of giants, as some had facetiously suggested, but for "strengthening the vitality of infants born before term or congenitally feeble."[5] He interpreted the incubator, in other words, as a means of promoting "vitality" rather than simply preventing heat loss. And in the nineteenth century, heat was but one component of vitality. Whereas temperature represented a concept that could be quantified and verified through the use of a thermometer, vitality connoted the mysterious forces that made life itself possible. Even Tarnier, it should be recalled, had not completely distinguished the two and often set the temperature of his incubators according to the infant's color, behavior, and heart rate as well as its body temperature.[6] But French obstetricians still emphasized temperature more than did their American counterparts. The resulting ambiguity had critical importance for the future evolution of the incubator, as Americans applied to the device their own notions of what factors promoted vitality.

Providing the energy for this transformation was a burgeoning wave of technological enthusiasm overtaking the United States at the time of the introduction of the incubator. It radiated into the medical profession from outside, through the values exemplified by the rise of professional inventors in American life. In 1876, Thomas Edison opened his lab at Menlo Park, and Alexander Graham Bell constructed the telephone; the two events symbolized the beginning of what historian Thomas Hughes has called the era of independent inventors.[7] Lasting until the rise of corporate research labs in the early twentieth century, this period witnessed an explosion of technological innovation. The United States approved more than twice as many patents in 1896 as in 1866, reaching by the end of the century a total exceeding that of Britain, France, and Germany combined.[8]

Without a secure corporate or governmental niche, professional inventors depended on revenues from patents and private investors for support; many, in fact, formed their own companies. Their independence from bureaucratic supervision contributed to a comparatively flexible and radical style of invention. Edison sought to develop new technological systems rather than improve existing ones, and Elmer Sperry attested a preference for problems promising 95 percent breakthroughs rather than 5 percent improvements.[9] Such figures characteristically distanced themselves from scientific theory and academic elitism. The idea of professional science remained insecure on American soil during these years and continued to arouse suspicions in a democratic culture. Though an inventor such as Edison might well call upon the services of a professional scientist, he rarely acknowledged doing so publically.[10] As a result, he and his counterparts became veritable folk heroes in popular culture, providing reassuring symbols that the self-made man was alive and well in an America whose pastoral Jeffersonian dreams were vaporizing in the furnace of industrialization.

Physicians both collaborated with inventors and became inventors themselves. Academic leaders, for example, typically called upon the services of a technical expert to design a new incubator, recalling the relationship between Tarnier and the Paris instrument maker Odile Martin. Outside of academic settings, physicians quite frequently provided the technical expertise themselves. Obstetricians in particular identified to a certain extent with surgeons, and surgery was fundamentally mechanical in nature. Contemporaries observed that American surgeons manifested an enthusiasm for new procedures that suggests a parallel with that of their fellow citizens for mechanical devices in general.[11] This enthusiasm for intervention often conflicted with the more conservative values of the academic leadership and in the case of obstetrics was frequently berated as meddlesome midwifery.[12] Regardless of such charges, physician-inventors exemplified a model of medical professionalism emphasizing common sense and mechanical ingenuity over scientific expertise and specialization.

To the extent that physician-inventors wrote at all, they often produced a type of professional article prominent in the late-nineteenth-century medical literature, the invention account.[13] One example was that of a crusty Minnesota physician, E. J. Brown, who described in 1892 his experience building an incubator when confronted with the unexpected delivery of a four-and-a-half-pound premature infant. Brown had just learned of the incubator through a recent article in the *Medical Record* but could find no precise description in his journals. He consequently decided to design "during an idle office hour" his own incubator out of a packing box, which he constructed during the evening out of materials from the family lumber room. A bracket lamp heating a pail of water allegedly required only minimal attention to maintain a uniform internal temperature at least 15–25°F degrees above that of the room. Brown's account, it should be noted, emphasized the virtue of improvisation over precision; he presented his device not as an ideal but as a simple but lifesaving device "which any clod-hopper ought to be able to construct in a few hours." Also notable was his focus on the process of invention rather than the care of the child, as if the former more or less provided for the latter. His account centered on the workshop rather than the nursery.[14]

Though it would be hard to prove whether such physicians were ultimately more interested in the machine than in the infant, they did often emphasize technological apparatus over technique. It was difficult for them to do otherwise, for in the early stages of the diffusion of incubator technology, knowledge of the apparatus itself spread faster than the techniques to operate it. Even in the French obstetric context, early accounts of the incubator in the 1880s to a great extent assumed, rather than articulated, the nursing techniques required to employ the device. Only when the incubator encountered setbacks did physicians such as Pierre Budin begin to make the role of the nurse and mother explicit.[15] In the meantime, medical inventors naturally concentrated their efforts on developing the incubator itself. Because much of the resulting innovation took place far from academic circles, medical theory shaped it at a distance. Under such circumstances, the relationship between theory and practice in the inventor's imagination is difficult to discern. But one possibility is suggested in Thomas Hughes' studies of American professional inventors: the use of metaphorical thinking.[16]

Metaphors played a pivotal role in the evolution of the incubator into a machine. This point is illustrated by the two central functional issues in its design, heating and ventilation. Both presented technical problems, often formidable. At the same time, both incorporated metaphors transforming the meaning of the incubator. In the process, the device became a symbol of the extension of medical control into the domain of the newborn. On an even more fundamental level, the very structure of the device suggested to some physicians a kind of scientifically fashioned artificial womb. The examination of this particular image offers a suitable starting point for interpreting the reinvention of the incubator as it crossed the Atlantic.

The Artificial Womb

The essential novelty of Tarnier's incubator lay in its design as a closed rather than an open system. Its other innovations, particularly the thermosiphon, were eliminated with the simplification of the apparatus into the Auvard model. As we have seen, the superiority of this device over the traditional expedient of the padded basket with hot water bottles was far from self-evident. Tarnier contended that, unlike a basket, a closed incubator warmed not only the infant's exterior but also the air it breathed; it heated the baby from inside as well as outside. Two factors limited the appeal of this argument to Americans. First, few appreciated the importance of precisely maintaining the infant's temperature within the narrow range of normal. The abstracts of Auvard's article that appeared in American journals nearly eliminated its original (and lengthy) section devoted to the effect of the incubator on individual temperature curves.[17] Second, traditional hygiene suggested that breathing cold air was actually beneficial so long as the skin was kept warm; it stimulated the lungs and protected them from infection.[18]

Yet whatever Americans might think of Tarnier's theoretical rationale, they had more difficulty brushing aside his statistics. To the extent that premature-infant survival furnished the standard for its justification, Tarnier had provided a powerful case that the incubator was in fact superior to the padded basket. It was true that he never personally compared the apparatus to its other principal European competitor, the *warmwänne* or double-jacketed warming tub. But the latter device had thus far attracted little attention in the United States, and even after its announcement by Credé in 1884 it remained very much overshadowed by the Tarnier incubator.[19] Indeed, the incubator almost literally swallowed it, for a number of physicians transformed the warming tub into a closed system. They developed incubators that were encased in a heated water jacket and yet covered from above as well. Though this type of design hampered observation, its defenders proposed a different advantage: it created a protective environment analogous to the womb.

The metaphor of the incubator as artificial womb drew on ideas about premature infant care that arose independently of Tarnier's invention. In 1882, not long after the *couveuse* was installed at the Maternité, the gynecologist Franz Winckel at the Dresden Royal Maternity Hospital published an account of his own apparatus for warming premature infants. His "permanent water bath" consisted of a metal tub actually designed to suspend the infant in a pool of warm water. Only the baby's head protruded though a collar into the outside world. The device was terribly difficult to use; the danger of the infant's head slipping into the tub required constant vigilance, and waste matter in the water posed nettlesome hygienic concerns. Even Winckel tried it on only ten cases, with few imitators following.[20] But the device was nonetheless superior to the *couveuse* in warming the extremities of the infant and maintaining its core temperature, as even Auvard admitted after his own experiments.[21] And it seemed to provide an environment more analogous to the womb. The unclothed infant was suspended in warm water, just as it would have remained in amniotic fluid had its birth been delayed until term.[22] The idea of imitating the womb was a powerful idea, one which American physicians soon developed as a rationale for the closed incubator.

The artificial uterus motif soon reappeared in one of the earliest original American incubators, the Rotch-Putnam model. First unveiled in 1893 at the World's Columbian Exposition in Chicago, this device was notable in being the product of a remarkable collaboration of two individuals, an academic physician and an inventor.[23] With such a divided parentage, it manifested a closer relationship to academic medicine than did many of the incubators produced by solitary physician-inventors. Thomas Morgan Rotch, the first professor of pediatrics at Harvard, was perhaps the best-known American authority on the scientific feeding of infants using the principles of chem-

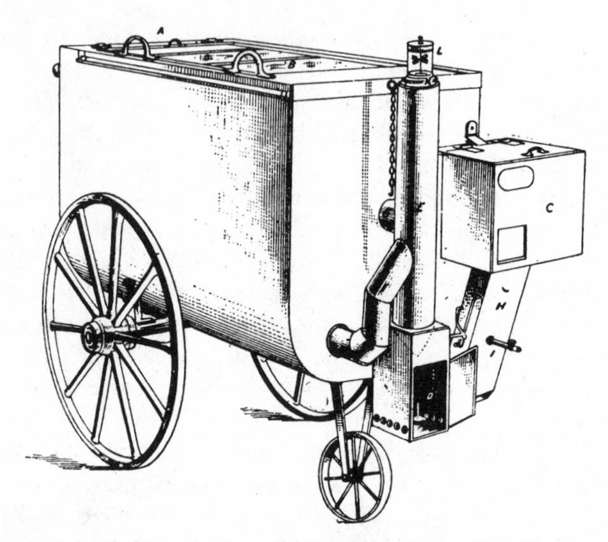

Rotch-Putnam brooder. *A*, scales for weighing infant; *B*, glass lid of incubator; *C*, fresh air apparatus, (clock work with fan); *D*, lamp for heating water jacket; *E*, Chimney; *F*, return flue from heating flues; *G*, return fresh-air flue; *H*, entrance for fresh air; *I*, connection for oxygen supply; *J*, mixing valve; *K*, ventilating exit; *L*, anemometer. *Source: Transactions of the American Pediatric Society* 5 (1893): 45.

istry.[24] He was also one of the first pediatricians to take interest in premature infants. Rotch commissioned a Boston technical expert, John Pickering Putnam, to develop the incubator from his design.

The novelty of Rotch's approach began with his criticism of earlier incubator designs for having addressed only one variable underlying infant vitality, that of temperature.[25] From his perspective, the advantage of a closed incubator over a padded basket was not so much as a more efficient agent to prevent heat loss but as a means of protecting the infant from the outside world into which it had been prematurely thrust. The closed incubator acted, first of all, to isolate the baby from dust and microorganisms and thereby protect it from infection; Rotch insisted that only filtered air enter the device. But he envisioned bacteria as but one threat to the infant's precarious vitality. Excessive stimulation, whether through touch, light, or sound, constituted another. "The premature infant," Rotch asserted, "should, so far as is possible, be restored to the condition that it has been forced out of—namely a condition of darkness, silence, and warmth."[26] Rotch's incubator was encased in a double-jacketed wall of copper containing a layer of heated water, reminiscent of Credé's *warmwänne.* But it was sealed from above, as well, and featured only a single small window that in turn was often covered by a black cloth. Vital signs such as temperature and respiratory rate were to be taken only with great care. In fact, Rotch to some extent replaced the thermometer with a new standard for monitoring the infant's progress: its daily weight. His incubator actually featured a self-enclosed scale to allow this variable to be monitored without human contact.[27] The Rotch incubator provided an artificial technological environment from which the infant almost never had to be removed.

Rotch and Putnam were far from alone in developing the intrauterine metaphor and in the accompanying notion that premature infants had a limited supply of vital energy that could be depleted by excessive handling. Perhaps the fact that premature infants spent so much of their time asleep suggested that too much wakeful activity could overwhelm their fragile nervous systems. Numerous authors, some of whom cited Rotch specifically, insisted that the incubator should simulate the in utero environment as far as possible. Within the uterus, wrote a practitioner in 1901, the infant "rests in a warm chamber, bathed, as we may say, on every side by a water bed, while the body-wall of the mother shields it from external violence. No ray of light or sound from the external world disturbs its delightful slumber."[28] Such descriptions of the uterus in terms of a "water bed" shielded by the mother's "body wall" provided a natural justification for the water-jacketed incubator. The layer of warm water within its walls underlined the role of the device as a protective barrier.

On another level, the idea of the incubator as artificial womb implied a potential challenge to the mother. One would not want to overstate this implication; Rotch certainly did not want to exclude the mother from the baby's care, as exemplified by the fact that he set his incubator on wheels to allow it to be used at her side in the home.[29] The idea of minimal stimulation nonetheless imposed at least a symbolic barrier between mother and infant. Rotch's argument here recalled his justification of artificial formula, which he presented as superior to breast milk for premature infants because it did not vary on account of "emotional causes."[30] Here he tapped into a widespread belief that "nerves" frequently disrupted breast-feeding and induced colicky symptoms in the child.[31] Science, he argued, could imitate and improve on nature; it could copy the desirable elements of breast milk and save the infant from the harmful influences of its own mother.

There was a fine irony here, for modern life, and modern technology in particular, were commonly understood as the underlying reasons why mothers suffered from "nerves" in the first place. The late nineteenth century witnessed a widespread preoccupation with nervousness as the preeminent American psychological affliction. Most famously, the neurologist George

Beard popularized the term *neurasthenia* to encompass a wide-ranging spectrum of functional disturbances, which he envisioned to be the consequences of exhausted nervous energy. Underlying the condition were the rapidly changing demands of American life in the Gilded Age, out of which Beard particularly singled out the changes exemplified by the telegraph, railroad, watches and clocks, noisy appliances, and the rising educational levels sought by women.[32] The peculiar susceptibility of women is worth emphasizing; these were the years in which the first generation of college-educated women was coming of age, in which the cult of domesticity was giving way to assertions of responsibility beyond the home. Education was commonly feared to exhaust the delicate temperament of the mother as well as her children.[33] Indeed, nervous exhaustion rendered the mother a potential threat to her child beyond pregnancy throughout the period of nursing. Leading pediatric writers warned that such innocuous practices as rocking and exciting young infants predisposed them to nervous afflictions later in life.[34] Premature infants, having the weakest and most sensitive reservoirs of vital energy, logically seemed most vulnerable to excessive handling. The incubator thereby became a kind of rest cure, a refuge from the stimulation of modern life.

The metaphor of the incubator as a refuge analogous to the womb underwrote a conservative therapeutic philosophy of minimal stimulation that would later dominate American premature infant care far into the twentieth century. Its rise was associated particularly with the career of the Chicago pediatrician Julius Hess, the foremost American premature infant advocate of the 1920s and 1930s. Hess's incubator consisted of an electric version of Credé's *warmwänne* with a hood that made little provision for observing (or manipulating) the infant within. His emphasis on conserving the strength of the infant preserved an essentially vitalistic nineteenth-century understanding of prematurity.[35] Yet it was not the only theme that introduced a symbolic tension between the infant and mother into the interpretation of the incubator. The issue of automaticity heightened the issue still further.

The Mechanical Nurse

Given that the original goal of the incubator was to warm the infant, it is perhaps surprising how much uncertainty persisted over just what this task required. As we have seen, the French maintained the interior of the device anywhere between 79° and 95°F, depending on the particular physician and infant. Budin recommended lower temperatures than did Tarnier, presumably reflecting the fact that he dealt mostly with larger premature babies.[36] And although the measurement of rectal temperature in theory provided a standard by which to set the incubator for a given infant, it too had to be qualified. Nurses took into account not simply the infant's temperature, which they typically measured only twice a day, but its level of activity and irritability.[37] There was further disagreement over whether physicians should always strive to keep a newborn infant's temperature within the adult range of normal (close to 98.6°F) or tolerate lower initial temperatures as long as normal values were reached within several days.[38] Fundamental questions regarding the goal of heat therapy thus remained unanswered throughout the early history of the incubator. Indeed, not until the 1950s did careful experimental studies create a consensus in favor of precisely maintaining the premature newborn's temperature within the adult's normal range.[39]

But if medical inventors disagreed on how warm to keep an incubator, they rarely challenged the proposition that its interior temperature needed to be stable. They therefore sought to incorporate mechanisms to make the machine more self-regulating. Such a strategy diverged fundamentally from that of the Paris obstetric tradition, which had quickly abandoned thermostats and elaborate heating systems. The propensity of the original thermosiphon within the Tarnier-Martin incubator to overheat quickly led to its abandonment. Tarnier's response was to simplify the device by restoring the tried-and-true method of hot water bottles monitored by a diligent nurse.[40] In other words, he sought a solution in the hospital's personnel rather than the machine proper. Budin took this approach a step further by finding a caretaker that he believed to be more reliable than the nurse, the infant's own mother.[41] Medical inventors, in contrast, sought to have the machine take over these functions by making it more automatic and free from attention.

The continuing analogy with poultry incubators encouraged this trend toward automaticity. The chicken incubator not only provided the immediate inspiration for Tarnier's invention; it influenced his imitators, as well. Although the origin of these devices dated back to at least the seventeenth century in Europe, they were undergoing a period of rapid evolution during the late 1800s. At this time, the United States alone sometimes issued as many as fifteen patents annually for new incubators and brooders.[42] Many were considerably more advanced than infant incubators, featuring heating systems employing steam radiators or even, by 1895, electric heating sources.[43] They constituted a pathway by which technological transfer could bypass the level of professional medicine.

Unfortunately, for this very reason, this route of innovation left few traces in the historical record. Inventors, by and large, were less likely than physicians to record their experience.[44] Only scattered references remain. A *New York Times* article in 1894 described how one such inventor, W. G. Robinson, decided to extend the concept of the chicken incubator to other problems. His first project, a laboratory germ incubator, was so successful that he soon opened a shop and began effectively competing with foreign suppliers. Robinson then went on to develop through a series of experiments an incubator for premature infants. The article focused on the inventor, rather than the physician or nurse, and presented the device as a dramatic innovation, its success illustrated by the rescue of an infant born three months prematurely who "is today a bouncing boy."[45] Other poultry incubator manufacturers attempted, like Robinson, to modify their devices for infants. The manufacturer George Stahl, of Quincy, Illinois, advertised an "infant nursery" for sixty dollars in 1895 as being identical in principal to his successful Excelsior incubator for eggs. Most striking was its provision of a thermostatic regulator, which enabled the device to be completely automatic, maintaining a constant temperature with minimal attention.[46]

The thermostat represented the greatest technical challenge for professional inventors interested in the incubator. The first practical thermostats emerged in the mid-nineteenth century, although there are sketchy accounts of precedents as far back as the 1600s. American inventors, in keeping with their traditional attraction to labor-saving devices, played a major (though not exclusive) role in their evolution.[47] Some designs, such as that employed by Budin's incubator of the early 1880s, relied on the expansion of mercury with heat. Others operated on the principal of the unequal expansion of metals, causing a bimetallic coil or strip to bend at a specific temperature.[48] The poultry incubator provided one of the first examples of their practical application. The British inventor Charles Edward Hearson developed one such thermostat as part of his popular Champion incubator for eggs in the early 1880s. He then transformed the apparatus in 1884 into a sophisticated incubator for premature infants. When the temperature inside Hearson's device exceeded a chosen level, the thermostat (consisting of a metal capsule that contained blotting paper soaked with petroleum) expanded, manipulating a system of levers that lifted a damper allowing heat from the gas power source to escape. Although Hearson's infant incubator was promptly acquired by the London Lying-in Hospital and imitated in France, the extent of its popularity is uncertain. He does appear to have been a pivotal figure in linking the thermostat and poultry incubator, setting off what became a vast industry in Britain.[49]

Incubator thermostats could be quite reliable in certain settings but were often found deficient in actual practice. Physicians frequently complained of their erratic behavior.[50] Yet in this case, the exceptions are important. There were isolated examples of well-known physicians who claimed to have developed trustworthy thermostats—most notably Joseph B. DeLee of Chicago, an obstetrician who was to become one of the most zealous proponents of the incubator in the United States.[51] DeLee, in fact, was the only other major figure within academic medicine besides Thomas Rotch to play the role of inventor as well as investigator with respect to the incubator before 1914. DeLee modeled his thermostat on those of local poultry incubators, and indeed its design (a biconcave expanding disk containing ethyl chloride) was similar to that of Hearson in England.[52] His testimony suggests that although the problem of the thermostat may have posed a major obstacle to physicians without technical expertise, it was not insurmountable. For most physicians, the problem seems to have been that thermostats performed well initially but were prone to break down without proper care and maintenance.

The experience with thermostats in the poultry industry provide an interesting comparison in this regard. Technical literature cited public demonstrations in which these devices maintained an internal temperature within half a degree for forty-eight hours, while manufacturers claimed similar results for periods as long as two weeks.[53] Farmers, on the other hand, sent letters to agricultural magazines denouncing egg incubators as unreliable.[54] The most reasonable way to reconcile these accounts is to suggest that the state of art in the late 1800s was sufficiently advanced to solve the problem of making a reliable thermostat, but not one that could survive what might be called a technologically unfriendly environment.

Further charging the controversy over the thermostat was the question of whether it was even desirable. The thermostat may have been laudable as a labor-saving innovation, but in this case the labor spared was that of the mother or her domestic nurse. The interests of the two were not synonymous. By the late 1800s, increasing numbers of middle-class mothers in urban areas were able to hire an infant nurse (or "nursemaid") to assist with infant care. Although hospital-trained professional nurses were also appearing by this time, the quintessential nursemaid until the end of the century remained an older woman schooled by experience rather than formal training. Trying to fit such an employee into the role of domestic servant could produce considerable tension. Given the frustrations that American women often expressed with finding "good help" in a socially mobile society, one might expect a technology that replaced the nurse to be attractive.[55] But entrusting an infant to a machine fundamentally challenged the ideal of the devoted and sacrificing mother.

Medical instrument makers were less eager than poultry incubator manufacturers to develop a truly automatic incubator. The Chicago manufacturer Truax, Green, and Company offered a compromise featuring a thermostat that merely activated a warning bell rather than directly controlling the heat supply. One physician defended this model in terms of both technical utility and maternal responsibility. "Automatic devices," S. Marx wrote, "while very convenient, are liable to derangement, and the attendant is relieved of the necessity of constant vigilance."[56] The thermostat threatened to undermine the role of the nurse. The rhetoric used by promoters of automatic incubators in the late 1800s betrayed a similar theme. Hearson intriguingly called his device a "thermostatic nurse," just as the first English report of the Tarnier-Martin incubator had called it a "mechanical nurse."[57]

Of all the thermostatic models that appeared in the late 1800s, one deserves to be singled out: the Lion incubator, patented in 1889. Ironically, it was again France, the country whose obstetric leaders remained so thoroughly wedded to the traditional *couveuse,* that produced the most popular automatic incubator of the early twentieth century. The Lion incubator, however, emerged outside of Paris, far away from the academic obstetric tradition associated so strongly with that city. Though little is known of its creator, Dr. Alexandre Lion, it is intriguing that he was the son of a professional inventor. Like Hearson in England, Lion first appears to have developed his large, thermostatically controlled incubator for chicken eggs. He then went on to test the device successfully on a premature infant of six months' gestation.[58] Interestingly, the justification of the device by Lion's admirers suggested why a French physician outside of Paris might take an approach similar to that of American and British inventors. The problem with the Tarnier incubator, they asserted, was that it required the attendance of experienced personnel who were difficult to locate outside of the Paris maternities.[59] The complexity of the Lion incubator was intended to free it from such dependency. From the beginning, it aspired not merely to extend but to replace the functions of the nurse. As a result, Lion became an agent of technological transfer rather than another example of the French obstetric tradition.

The Lion incubator represented the climax of incubator design at the turn of the century. To begin with, the device presented an imposing sight visually. Not only was the infant's chamber enlarged, but the entire apparatus was situated on sturdy metal legs that had the incidental effect of rendering its overall size comparable to that of the nurse. When displayed in public, the streamlined metal and glass surfaces of the new incubator, polished to a silvery finish, suggested the aesthetic of industrial rather than domestic art.[60] The showy exterior encased a heating system that was both complex and self-regulating. A gas or oil lamp heated an external boiler, causing its water to circulate through a spiral radiation coil before returning. Into this system Lion incorporated an electric thermostat regulating the heat

Lion incubators in incubator station, Chicago Lying-in Hospital, 1901. *Source:* Chicago Lying-in Hospital, *Annual Report* (Chicago: S. Ettlinger, 1901–2), 16.

source through a series of levers. His supporters claimed this thermostat to be reliable for hours and even days, effectively relieving the nurse from the need for constant attendance, just as the heating system freed her from having to replace the water every two to three hours.[61] The machine could now warm the infant and monitor its environment. One admiring physician remarked that the infant born prematurely was no longer likely to miss his "umbilical circulation" thanks to the "perfection of modern industry" embodied in the Lion model.[62]

At the same time, Lion's model represented the culmination of another aspect of incubator design. It included a sophisticated ventilation system enabling it to acquire its air supply from outside. This feature probably commended it to American physicians far more than did the thermostat, for they had already begun to reconceptualize the critical function of the incubator in terms of ventilation rather than heat.

From Heating to Ventilation

The most profound transformation overtaking the incubator as it left Paris was a shift from heating to ventilation as the central issue in its design. All of the incubators employed in the French obstetric tradition circulated air by convection. In other words, they simply allowed the air within the incubator to rise by virtue of its own warmth, first around the infant and then out through the exit holes. Convection was most effective when the incubator was kept in a room much cooler than its interior; in such settings it generated sufficient force to cause a spinning anemometer to whirl, disperse a bed of feathers, or blow out a lit candle.[63] These conditions were commonplace in Paris maternity hospitals, where temperatures could drop close to 50°F in the winter and, thanks to the insulation provided by their massive exterior walls, rarely rose above 80°F even during the worst of the summer.[64] Tarnier, Budin, and their followers thus rarely, if ever, expressed concern over ventilation.[65] In contrast, physicians in Britain and the United States singled out the problem almost immediately as a potential shortcoming of the apparatus.[66]

Physicians spoke of ventilation on at least three different levels, often in combination. The first most closely anticipated the common use of the term today: the renewal of air required to prevent suffocation within a closed space. On occasion, physicians who noticed acute symptoms of respiratory distress or cyanosis in a premature infant blamed faulty ventilation.[67] Even some defenders of the apparatus admitted that convection alone was inadequate on hot summer days, and they propped open the lid under such conditions. Yet this criticism alone should have precluded the use of the device only in unusually warm conditions, not during the remainder of the year when the wards were generally cool (and the provision of warmth for an infant most problematic).[68] The fact that ventilation came to be employed as a general criticism of the Tarnier incubator suggests that the term implied more than just moving air. Indeed, most physicians discussed it in relation to preventing infection rather than suffocation.

Ventilation was closely linked to the problem of preventing infectious agents from invading the space within the incubator. In France, it will be recalled, the creation of *services des débiles* during the 1890s admitting babies born outside of the hospital resulted in devastating outbreaks of bronchopneumonia. Controversy erupted over whether the incubator itself was at fault. Dividing the two sides was the question of whether these infections were spread primarily by contact or through the air. If the former were true, it would be possible to follow Budin's strategy of preventing infection through the application of antiseptic agents and isolation techniques on the ward.[69] But the prevention of infection from the air was far more problematic—particularly from the standpoint of ventilation. A filter of cotton or gauze might offer some theoretical protection from airborne microorganisms, but at the cost of impeding air flow. The finer the filter, the greater the problem of ventilation.[70]

Yet ventilation went beyond the questions of air movement and preventing airborne infection; it could imply a third level of meaning expressed in the idea of fresh versus stagnant air. It carried, one might say, cultural as well as scientific meanings. In the United States and Britain, these beliefs were closely linked to the powerful sanitarian tradition so prominent in both countries. This tradition, itself shaped by romantic ideals of pastoral life and disdain of the city, offered an explanatory framework rendering disease a symptom and symbol of the devastating effects of industrialization in those countries. In particular, its leaders refined and popularized the ancient notion linking disease to invisible atmospheric poisons produced in conditions of squalor. Ventilation thus became one of the key environmental determinants of health. The sweltering heat and confined conditions of New York tenements robbed their residents of vitality just as assuredly as they induced putrefaction in the food those same occupants ate. The remedy for diseases associated with crowding and poor ventilation, such as tuberculosis, logically seemed to be the restoration of fresh air. And in fact, sanitarian programs had remarkable success in reducing urban disease well before the germ theory by simply getting rid of filth.[71] The linkage between fresh air and health in the nineteenth century became almost a matter of common sense in the United States.

One would not want to draw too sharp a line between the cultural and scientific understandings of ventilation, for few physicians saw any significant conflict between the two. Perhaps the most influential scientific attempt to rationalize ventilation during the second half of the nineteenth century was that of Max von Pettenkoffer in Germany, who in the early 1860s attributed the deleterious effects of poor ventilation not to carbon dioxide, as earlier theories had suggested, but to organic poisons exhaled by the body. These "morbific matters" or "crowd poisons" depleted the body of its resistance to infection.[72] This kind of reasoning withstood the discovery of bacteria as the immediate cause of infection in the 1870s and 1880s.[73] Physicians continued to invoke ventilation to explain why only a fraction of those exposed to a particular pathogen, such as tuberculosis, actually succumbed to the illness. They reasoned that the atmosphere itself had an effect on the body's resistance; foul air, though no longer thought to be the direct cause of sickness, might still represent the key element in predisposing the body to infection.[74] And since resistance to infection was closely linked to the older idea of vitality, the prospect of placing an infant in the boxlike incubator was troublesome, indeed.

Again, the emphasis on ventilation suggested a new metaphor for the incubator as an artificial home. One prominent example of a physician who employed this language was none other than Thomas Rotch, who called his 1893 incubator design a brooder to emphasize that it served as "not merely a receptacle but an actual habitation for premature infants" where they might reside for weeks or even months.[75] Significantly, his technical associate was an expert not in poultry incubators but in domestic engineering. Indeed, John Pickering Putnam had written a book that analyzed how the placement of fireplaces and windows influenced air currents in the American home. His brooder incorporated an elaborate system of ventilating flues, allowing heated and fresh air to be mixed in exact proportions. The apparatus also included a small fan to propel air through the system when convection alone appeared to be inadequate. The provision of a quiet fan before the widespread availability of electricity represented still another technical challenge. Rotch hoped to power the device eventually through an electric battery but in the meantime fell back on Putnam's expedient of a clockwork fan.[76]

Although the Rotch-Putnam brooder thus differed from the Tarnier *couveuse* in having a mechanical means to propel air through its interior, it still made no provision for acquiring its air supply from a pure source. Rotch and Putnam, in fact, mounted the device on wheels and intended it to be a domestic technology, one that would be used in the home rather than the hospital.[77] Its lack of provision for obtaining outside air limited its appeal when challenged by another approach intended for institutional use, that of connecting the incubator to an external forced-air ventilation system. The Lion incubator exemplified this alternative most elegantly.

Lion created a device intended not merely to heat but to "process" the atmosphere breathed by the infant.[78] This concept meant obtaining air from outside the building, propelling it by means of a system of conduit pipes to the individual incubators, and then filtering, humidifying, and finally warming it. Air exited through a separate system of exhaust pipes. Three instruments—a thermometer, hygrometer, and anemometer—enabled the attendant to monitor the internal environment of the device. In paying such extraordinary attention to the source and quality of the incubator's atmosphere, Lion clearly expressed his belief in the value of outside air for premature infants. Less obvious is why he valued it: whether because of a presumed stimulating effect on vitality, the atmospheric poison theory of ventilation, or as a means simply to avoid exposure to airborne organisms. Whatever Lion's own rationale, his solution may well have appealed to more popular concepts of ventilation than did Tarnier's boxlike predecessor. A British in-

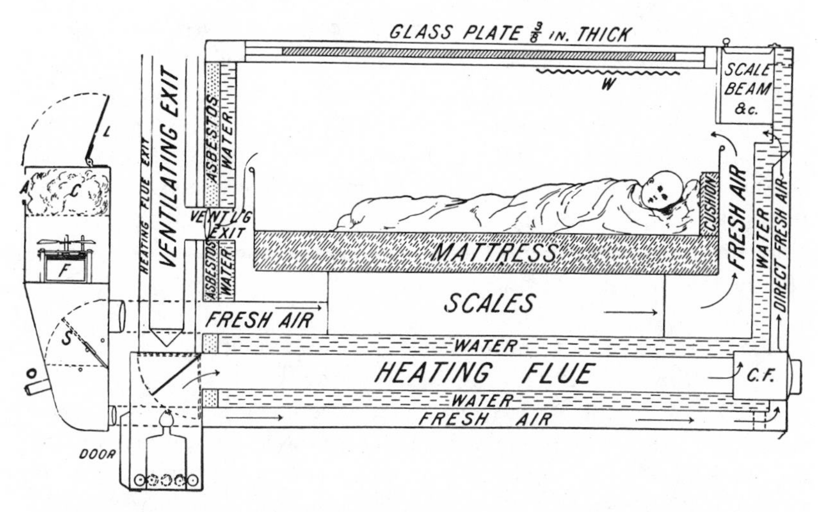

Rotch-Putnam brooder: cross section *L,* lid of fresh-air box, open; *A,* entrance of fresh air; *C,* cotton, resting on wire shelf above clockwork; *F,* clockwork and fan; *S,* valve regulating hot and cold fresh air; *O,* pipe for oxygen attachment; *C.F.,* cleaning-flue; *Door,* door to lamp-box; *W,* wire frame to protect against breakage of lid. *Source: Transactions of the American Pediatric Society* 5 (1893): 45.

terviewer in 1896 tried to rephrase the function of Lion's machines in terms meaningful to his readers, asking the physician if his incubator "by artificial means, stimulates the baby's vitality, and stimulates the weak organs of his body?" Lion answered in the affirmative, suggesting at least that if he did not rationalize the operation of his incubator in such terms himself, he did not contradict those who did.[79] The ambiguity underlying the working of Lion's apparatus represented on one level a strength, for it matched the complexity and inconsistency that characterized the contemporary understanding of ventilation.

At the same time, the various meanings connoted by the popular usage of ventilation could also suggest a conflict between scientific technology and domestic hygiene. There were real questions of whether the benefits of fresh air could ever be administered satisfactorily through any artificial means of ventilation. Such concerns were hardly new; Florence Nightingale, for example, had charged in 1863 that artificial ventilation by means of machines was "not in accordance with Nature's method of providing fresh air," since it did not vary in temperature and humidity according to the cycle of the day and seasons. Patients needed fresh air, warmed by radiation from the fireplace, rather than "warm water in iron pipes."[80] It was, of course, just such a system of artificial ventilation that incubator enthusiasts were trying to construct. The idealization of "natural" ventilation remained a powerful theme well into the early twentieth century and gained still further influence in the wake of the antituberculosis campaigns.[81] The image of the incubator as artificial home joined those of the artificial nurse and artificial uterus to challenge the traditions of domestic hygiene employed by mothers in caring for their newborns. To an even greater extent than the other metaphors, it provided fuel for the controversy that would engulf the device as it became established in the United States.

By 1900, the American term *incubator* had come to include a bewildering array of devices for premature infants. They ranged from Auvard *couveuses* heated by water bottles, to double-jacketed warming tubs, to the extremes of complexity represented by the Rotch-Putnam brooder and the Lion incubator. Indeed, it is far from clear whether inventors had anything in common more than the barest assumptions regarding just what an incubator was supposed to do. Whereas Tarnier had seen the device as a warming apparatus, Rotch saw it as an artificial environment that relegated temperature to a secondary role. And while Budin had promoted the virtue of simple incubators to reinforce the primacy of the mother, Lion and Hearson introduced thermostats to free it of such dependence. Contemporaries in different settings and contexts thus interpreted the incubator in quite distinct ways. One might even argue that it no longer made sense to use a single term to encompass all of these varied devices.

The evolution of the incubator illustrates how a given artifact may be perceived quite differently by individuals in different social contexts. Sociologists of technology have called this phenomenon the interpretive flexibility of artifacts.[82] Technology no more defines its own meaning than does a Shakespeare play; to some extent, it always must be interpreted. In this example, physicians and inventors failed to agree on how to improve the incubator because they disagreed on its purpose. Without such consensus, they drew from an assortment of metaphors and technological analogies to refine the device. The result was a branching pattern of evolution. The incubator did not emerge in response to an objective problem posed by an entity called "the premature infant." Instead, individuals responded to their own notions of what such infants required, drawing from a reservoir of artifacts, concepts, and models found in their own environment. A given incubator thus came to incorporate a certain vision of premature infant care as well as the hardware that made such care possible.

This theme of interpretive flexibility characterizing the incubator can serve as a means of illuminating American attitudes toward the newborn in the early twentieth century. For technology involves more than just gadgets and mechanisms; it captures meanings and values as well. As the incubator became more complex mechanically, it acquired a threatening aspect. The metaphors of the artificial uterus, nurse, and environment symbolized the effort of science to imitate and improve upon nature. In the process, the incubator came to embody a challenge to the traditional notion that the key to raising a weak or premature infant lay in entrusting it to the mother and a healthy home environment. It gained the potential to become a lightening rod for more widespread anxieties regarding whether a sick newborn should be regarded as the responsibility of the physician or of the mother.

CHAPTER FIVE

Propaganda for the Preemies

Having been reinvented in the image of science, the incubator stood poised to take the United States by storm in 1900. Indeed, the next several years gave rise to perhaps the most remarkable phenomenon of its early history, the incubator show. Public fairs and amusement parks began setting up concessions featuring live "incubator babies" on the Midway. The concept originated in Europe and then reached new extremes and popularity in the United States. At its center was the physician and showman Martin Couney, a self-designated missionary for the incubator crusade. Couney had been involved in earlier expositions in Germany, France, and Britain but acquired most of his fame after immigrating to the United States in 1903. His incubator baby show at Coney Island lasted until 1943, becoming the longest lived of all the resort's attractions before the Second World War. He also produced widely publicized exhibits in the 1933 Chicago Century of Progress Exposition and the 1939 New York World's Fair, both of which helped build him a reputation as the preeminent, if unconventional, pioneer of American neonatology.

For many American neonatologists today, one of the most intriguing historical questions pertaining to their specialty is whether Couney and the incubator shows represented a public spectacle or a critical link between French and American neonatal medicine.[1] Couney himself encouraged others to see his career in these terms, particularly when he became friends later in his life with the Chicago pediatrician traditionally considered to be the father of American neonatology, Julius Hess. In a 1939 interview, he claimed to have trained under Pierre Budin, and by the time of his death in 1950 his role as intermediary between Budin and Hess had become part of his obituary record.[2] Yet the story is not so simple as Couney would suggest. His connections with Budin were more superficial, and those with Hess more complex, than apparent at first sight. It is possible that Couney's career was the most spectacular manifestation of the process of technological transfer, not the most influential.

Examining the use of incubators in expositions holds interest for the historian from another standpoint. Within the setting of the fairs, the incubator engaged the public as well as the medical profession. Nearly all writing on the incubator appearing in the lay press during the early twentieth century was linked in some way to the shows. Such accounts offered at least glimpses of how nonphysicians understood the incubator and the patients it treated. The fairs also provided opportunities for physicians located away from the East Coast academic elite to use sophisticated neonatal technology. It can be argued, in fact, that one such physician, St. Louis's John Zahorsky, deserves more attention than Couney as the individual who led the most energetic attempt to use the fairs to excite the interest of his profession in the premature newborn. Yet he is not remembered as a pioneer of neonatology, in spite of having produced the finest American monograph on prematurity of his day. His example is illuminating, for it reflects the dilemmas posed by the uncertain legitimacy of the incubator shows in the early twentieth century. The story is best approached by considering the phenomenon's origins and its early evolution before evaluating the careers of Couney and Zahorsky.

Technological Idealism: A New Vision of Neonatal Care

The incubator show did not begin on the Midway. It evolved through the attempts of Alexandre Lion of France to find a means to promote his remarkably complicated and expensive incubator. The Lion model dominated the shows throughout their history, an essential point to remember in attempting to understand the early appeal of these exhibitions. In 1900, the Lion incubator represented what might today be called high-tech medicine. Modern neonatologists would likely consider it the most advanced of the early incubators, certainly to the extent that it incorporated automatic heat regulation and forced-air ventilation. Admittedly, its emphasis on obtaining fresh, outdoor air did not become a permanent feature of incubator design. The apparatus nonetheless exuded an overall impression of technical sophistication that still seems to represent a major step beyond the Tarnier *couveuse.*

For contemporary admirers, the appeal of the Lion incubator was not simply that it was complex but that it promised to transcend the shortcomings of its institutional environment. Lion had specifically created it for use in locales lacking access to the level of nursing care provided by the Paris Maternité.[3] One may contrast this approach with the aphorism of the pediatrician Victor Hutinel that the *couveuse* was only as safe as its environment.[4] Certainly Lion's emphasis was very different than that of his contemporary Pierre Budin, who placed responsibility for temperature and infection control on the incubator's attendants. Budin admittedly did attempt to install some of Lion's devices in the Maternité but gave up after finding the hospital's gas supply too erratic at night for them to be reliable.[5] Later innovators would introduce a gas regulator to circumvent this problem, again enabling the device to overcome the deficiencies of its setting. But Budin, given his overall emphasis of focus on the mother and nurse rather than on technol-

Reunion of "graduates" of Maternité Lion, Nice, France, 1894. *Source: Strand Magazine* 12 (1896): 775.

ogy, was not the physician to do this. His perspective was very different from that of Lion, who deliberately sought to isolate the incubator from its original context, to make it more self-contained and therefore capable of crossing between diverse institutional settings. Lion produced a spectacularly anomalous technology that did in fact generate great public enthusiasm. The problem lay in how to sustain its development. Sophisticated medical technology, in the 1890s as well as today, was expensive.[6]

Rather than adapt his incubator to an existing setting, Lion developed a new kind of institution to contain it. With the aid of local philanthropy and the municipal government, he created the Maternité Lion of Nice, France, in 1891.[7] In spite of its title, the Maternité was not a traditional hospital but a specialized institute to provide incubator care of a decidedly technological variety. It was also a charity, accepting infants from all social classes, yet without the stigma attached to the Paris maternity hospitals.[8] It quickly became a great success, and by 1894 it had produced enough "graduates" to stage a public reunion. Pictures of long-skirted mothers carrying their now-plump former incubator babies provided a powerful form of testimony going beyond statistical tables.[9] Not that statistics were lacking; the city's physician-general simultaneously published a study demonstrating a 72 percent survival rate among the 185 infants brought to the charity. So impressive were

Interior of Lion's Paris incubator *charité,* 1896. *Source: Strand Magazine* 12 (1896): 771.

these results—a speaker at the Academy of Medicine called them "reassuring, even surprising"—that they bestowed upon Lion a new level of credibility. Their appearance as depopulation anxiety increased in the mid-1890s fueled his reputation still further. The same speaker reviewing Lion's work in 1895 admonished his countrymen to follow his example: "Since the number of infants born is diminishing more and more, let us try at least to save and raise them for the fatherland."[10] Lion heeded the call.

In his efforts to introduce his incubator institutes to other cities, Lion struck upon the idea of opening them to the public for an admission fee. The formula turned out to be remarkably successful, resulting in the establishment of storefront incubator institutes in Paris, Bordeaux, Marseilles, and Lyons by 1896. For fifty centimes, the curious visitor could enter a working premature infant nursery situated, for example, amid Paris's hustling Boulevard Poissionère. Inside were eight separately ventilated incubators and a glass-enclosed feeding area. Lion called these institutions *oeuvres maternelles des couveuses d'enfants,* a term that he translated into English as "incubator charities."[11] Yet the admission of paying visitors rendered their similarity to other infant welfare stations in France increasingly tenuous.

These incubator stations symbolized just how much distance separated the approaches of Budin and Lion. The idea of using incubators to care for infants just a few steps away from the smells and sounds of Paris street life would have been unthinkable with the Tarnier *couveuse.* It was just the sort of challenge the Lion incubator was intended to meet. The very purpose of the apparatus was to protect its occupant from the outside world, to preserve it in a hermetically sealed environment sheltered from its surroundings. Matching this freedom from its environment was a corresponding freedom from dependence on trained nurses. Lion hired nurses for six-month shifts, implying that he valued not so much older women with years of experience as young nurses able to work intensively for short periods. There was a sense in which the incubator itself had become the real nurse; Lion even charged a monthly rental fee for the device equivalent to a nurse's salary. To be sure, the fact that Lion maintained a high (1:3) ratio of nurses to infants suggested that nursing care was far from superfluous.[12] The image projected to the public, however, centered foremost on the technology.

From these storefront incubator charities, it was but a short step to incubator baby shows in fairs and exhibitions. It was at this point that Martin Couney's involvement in the incubator crusade began in earnest. Unfortunately, virtually everything known about this controversial figure's early career derives from an interview he provided for the *New Yorker* in 1939. According to this account, Couney was born in the disputed territory of Alsace that had passed from France to Germany in 1871. He attended medical school in Germany and subsequently made his way to Paris with the express purpose of studying under Budin at the Maternité. Budin, Couney related, was looking for ways to interest the international community in the incubator and sent his Alsatian disciple to display the device at the Berlin Exposition of 1896. Couney decided that such an exhibit would be far more popular if it incorporated live premature infants. The director of Berlin's renowned Charité hospital, Rudolph Virchow, promptly loaned him six babies from the maternity ward, with the excuse that most were likely to die anyway. As it turned out, the show's success exceeded even Couney's expectations. Christened the *Kinderbrutanstalt* (literally, "child hatchery"), it became a public sensation, drawing over a hundred thousand visitors in two months and even inspiring music-hall songs. Still more astonishingly, all the infants survived.[13]

Thus began the most peculiar career of all the early incubator enthusiasts. By the time the Berlin exhibit had closed, the British showman Samuel Schenkein had recruited Couney to set up a comparable exhibit the following year to coincide with the celebration of Queen Victoria's Diamond Jubilee. Here Couney ran into more resistance. The London hospitals refused to loan him any infants, threatening cancellation of the entire show. Couney returned to Paris, where (assuming his colorful testimony to be factual) none other than Budin provided him with three washbaskets full of premature foundlings. The popularity of the resulting exhibit at Earl's Court in 1897 matched that of its predecessor. Struck by the popular interest stimulated by both shows, Martin Couney resolved to become a professional doctor-turned-showman.[14]

What Pierre Budin thought of the incubator show phenomenon is hard to say. On the one hand, Couney's explicit testimony suggests outright cooperation in the Berlin and London shows. The timing is plausible; these were the years in which Budin, as director of the Maternité's *pavillon des débiles,* perceived the incubator campaign to be on the defensive. The fact that Couney brought to London and his subsequent shows an experienced Maternité nurse, Madame Louise Recht, further substantiates his own account.[15] Nonetheless, the alliance between Budin and Couney must be qualified in two respects. First, the incubators Couney employed were those of Alexandre Lion, whose business provided the financial backing for the Berlin exhibit. Contemporary reports of Berlin in fact mentioned only Lion, suggesting that Couney was likely working under his auspices.[16] Second, there is no evidence that Budin further cooperated or corresponded with Couney after 1897. Certainly, the two men diverged with respect to their relative emphasis on technology versus the mother. Couney never shared his former mentor's disillusionment with the Lion and promoted a style of nursery far removed in spirit from the mother-infant unit that would become associated with Budin. Indeed, although it is understandable that Couney would want to be thought of as a disciple of Budin, the best-known French pioneer of neonatal care, in truth he stood more in the technological tradition of Lion.

Couney's very success attracted less scrupulous imitators whose own activities fueled a controversy that would eventually ensure Couney's physical isolation from Budin as well. The great popularity of the Earl's Court exhibit caught the eye of Barnum and Bailey's headquarters at Olympia, which soon created its own incubator baby show. By February 1898, *Lancet* had issued a scathing editorial cataloguing an entire list of self-styled incubator exhibits in settings ranging from the Royal Aquarium to the Agricultural Hall at Islington. "What connection is there," the editors fumed, "between this serious matter of saving human life and the bearded women, the dog-faced man, the elephants, the performing horses and pigs, and the clowns and the acrobats that constitute the chief attraction to Olympia?"[17] Though specifically excluded from the condemnation, Couney decided to search for a more hospitable environment. The following year, he first ventured across the Atlantic.

Couney arrived in the United States in 1898 on the crest of a growing wave of enthusiasm for the Lion incubator. Seen from American eyes, the sideshow aspect of the shows remained subordinated to their celebration of a new technology. Medical journals began to publish a second wave of articles on infant incubation, and the lay press soon followed. The instrument makers Paul Altmann of Germany and Kny-Scheerer of the United States (a branch of a German company) issued their own versions of the Lion in rapid succession.[18] The claims of the device to be automatic and free of its environment lay at the heart of its appeal. The British press repeatedly pointed out how little attention the device required, one commentator going so far as to suggest that it could be rented out in place of the nurse.[19] Medical journals lauded its ventilation system as well; in fact, the *Lancet* editorial condemning incubator shows spent as much time criticizing showmen for ignoring the principles of ventilation as it did deriding the carnival aspects of their concessions. It was not clear whether being located next to the bearded lady or having to breathe the same air as the leopards of Wombwell's menagerie represented the greater danger.[20] The most striking aspect of medical coverage of these shows was how rarely it was unfavorable. Even the Barnum and Bailey show was described in sober and scientific terms by an American medical journal.[21]

Given such favorable attention, it should come as little surprise that the incubator-show movement reached its apogee in the United States in the early 1900s. Martin Couney was perhaps its principal, but not its only, promoter. His first American exhibit took place in 1898 at the Trans-Mississippi Exposition in Omaha, a setting that was probably too isolated to excite much reaction nationally.[22] He then returned to Europe during the following two years, a time period for which almost nothing is known regarding his whereabouts beyond his participation (probably in conjunction with Lion) in the Paris Exposition Universelle of 1900.[23] Couney finally captured the American spotlight when he reappeared at one of the largest exhibitions of the turn of the century, the Buffalo Pan-American Exposition of 1901. Articles appeared in professional and popular journals ranging from the *New York Medical Journal* to *Scientific American* and *The Cosmopolitan*.[24] The exhibit became one of the most popular on the Midway. Struck by his success and the prospects afforded by a country whose frequent fairs and expositions promised numerous future opportunities, Couney immigrated to the United States in 1903. The following year, another group of incubator enthusiasts presented a major incubator show at the St. Louis Louisiana Purchase Exposition of 1904.[25] The incubator was entering the public as well as the professional arena.

But the public stage on which the incubator had arrived was certainly a peculiar one. What are we to make of these exhibits featuring complex technology and live infants on the Midway? As it turned out, their impact on professional opinion turned out to be as ephemeral as the fairs themselves. In many ways they became literally a sideshow, in relation to the development of professional opinion regarding premature infants in the United States. Nonetheless, they remain of interest for the way in which they illuminate attitudes toward premature infants outside of the medical profession.

Education and Entertainment

In a speech at the Buffalo Pan-American Exposition of 1901, President William McKinley affirmed the higher purposes of the American expositions becoming ever more popular in his day. "Expositions are the timekeepers of progress," he proclaimed; "They record the world's advancement. They stimulate the energy, enterprise, and intellect of the people and quicken human genius."[26] It was within this tradition of the exposition as a public educator that Couney claimed a place for his incubator shows. The first great American exposition, the Philadelphia Centennial Exhibition of 1876, had been first and foremost a celebration of technology. Its focal point consisted not of entertainers but of inventions: Bell's telephone, Edison's duplex telegraph, the typewriter, and the sewing machine.[27] Expositions provided inventors a base from which they could promote their new products across local, national, and cultural boundaries. A Tarnier *couveuse* displayed at the Paris 1889 Exposition Universelle inspired the Boston instrument makers Codman and Shurtleff to manufacture one of the first American incubators in the following year. Thomas Rotch, likewise, displayed his brooder at the second of the great American expositions, the 1893 World's Columbian Exposition in Chicago.[28] Such exhibits were aimed at manufacturers, not the general public. Couney's decision to incorporate live premature infants, on the other hand, aspired to reach a broader audience.

Entertainment frequently competed with education in international expositions, especially when displays featured human subjects. This generalization particularly applied to the United States, where fair entrepreneurs had to rely on gate receipts and concession revenues rather than the government support available in Europe.[29] Although the Philadelphia centennial exhibition attracted an unofficial shantytown of amusements outside its gates, the decision of the World's Columbian Exposition to create an amusement area within the fair boundaries proper constituted a turning point. So successful was the Chicago Midway Plaisance that its namesakes became permanent fixtures of subsequent world's fairs. Native and exotic villages lured crowds beyond the bounds of conventional respectability to peer at Dahomay savages and dog-eating Filipinos. Such exhibits had originated in France, ostensibly with the blessing of anthropologists, but by the end of the century they increasingly portrayed their subjects in an imperialistic and demeaning fashion.[30]

An uncertain line demarcated such spectacles from another form of amusement that became a staple of the Midway by the early twentieth century—the freak show. The public exhibition of human curiosities for profit grew into a thriving entertainment business in late-nineteenth-century

America. During these years, hundreds of freak shows traveled across the country, performing in a variety of milieus ranging from circus sideshows to dime museums in rented storefronts.[31] The extremes of human size, midgets and giants, exerted a special fascination for the crowds attracted by such exhibitions. P. T. Barnum, a master at clothing popular entertainment in the language of scientific and cultural uplift, promoted "smallest baby" contests at his American Museum in New York City years before Couney set foot in the United States.[32] Supposedly inviable premature newborns who survived to childhood were objects of popular interest and curiosity; in 1893, a set of three-month-old twins allegedly born after less than seven months' gestation were featured with their parents at Curio Hall in the Eden Musée of St. Joseph, Missouri.[33]

The incubator show in the international exposition context fell into an ambiguous niche between education and entertainment. The Buffalo exhibition of 1901 featured a far more elaborate incubator station than could be found in any American hospital. The exhibit's babies lived and slept in a dozen or so Lion incubators, from which wet nurses could take them upstairs by elevator to special feeding quarters. Couney, by all accounts, tried to minimize the sensational aspects of the show and hired lecturers to expound on the finer points of raising a premature infant. Barkers outside the exhibit promoted its educational value and lack of "unpleasant features."[34] Yet the fair's management refused to allow Couney a place within the technological section of the fair, relegating him instead to the Midway. There, across from the Japanese Village and the Scenic Railway, the incubator building forced what one observer called "a violent change of mental atmosphere" from its surrounding milieu.[35] Couney charged an admission fee as did all Midway concessionaires, launching what became not only an educational but also an enormously profitable enterprise that drew crowds rivaling any of the competition.

Other showmen provided some of the most meaningful testimony to its success. Edward Bayliss, the president of the Exposition Amusement Corporation and producer of the Great Fire exhibit at Dawson City, Alaska, in Buffalo, created his own incubator show three years later at another major fair, the Louisiana Purchase Exposition in St. Louis. Bayliss housed his exhibit in a spectacularly flamboyant building featuring ornate towers and an inner courtyard lined with columns. Staffed with a full complement of nurses and even resident physicians, it represented the culmination of the early incubator shows.[36] Yet Bayliss was not alone in recognizing Couney's popularity; the proprietors planning Coney Island's Luna Park invited him to set up a permanent show when the park opened in 1903. Couney accepted and left France to become a permanent resident of Long Island.[37]

To the extent that the incubator shows were educational, their promoters sought to deliver a message. On one level, they sought to link the device to the ideals of medical science, which for many Americans meant German medical science. "It is something more than an exhibit," one popular magazine account proclaimed; "It is an educator."[38] The exhibits provided thousands of visitors their first introduction to the incubator, as well as other aspects of the "scientific" management of prematurity. They also made tangible a futuristic image of technological medicine, of machines dwarfing their tiny charges, of science accomplishing the miraculous. A St. Louis fair bulletin proclaimed, "It is the nearest realization man has ever seen of the dream of the old philosopher who looked in vain for the creation of human life through the processes of his alchemy."[39] Numerous popular and professional magazines reported the exhibits, providing historical and medical background information that may well have been provided by Couney and his lecturers.

From these sources, Couney appears to have publicly claimed an 85 percent overall survival rate with the incubator.[40] Most accounts played up the German roots of the device, which became more plausible now that the instrument was manufactured by Paul Altmann, the instrument maker who had supplied Robert Koch's laboratory. One magazine article even presented a history of the device that excluded the French entirely, proceeding directly from Credé's warming tub to the creation of a Berlin Incubator Institute in the 1890s. Here was a message consonant with the Teutonic sympathies of American medical leaders. It encouraged the visitor to leave with "a feeling of thankfulness for that genius that constantly broadens the scope of man's loving labor for his fellows."[41]

Other writers interpreted the exhibits as offering different messages to men and women. An illustration of the Berlin incubator exhibit in the English pictorial magazine the *Graphic,* entitled "An Artificial Foster Mother," centered on the contrasting reactions of a man and a woman to two respective incubators. The man was shown standing back, subjecting the exhibit to scientific scrutiny; the woman was smiling and leaning forward, her sympathetic gaze directed not at the machine but at its tiny human occupant.[42] Arthur Brisbane, a writer who later became a featured columnist for the Hearst newspapers, wrote an essay in Buffalo expounding a message along analogous lines. "Men go to the Exposition to see and to think," he asserted as he launched a sustained (and rather forced) comparison of the untapped energy of the incubator baby and the power of the exhibition's power plant at Niagara Falls. Although he confessed that most men were likely to be more impressed by the falls, the baby's brain represented a spiritual force that would launch works long outliving the waterfall. The main thrust of Brisbane's metaphor was an appeal to regard the premature infant as a resource,

"An Artificial Foster-Mother: Baby Incubators at the Berlin Exposition," 1896. *Source: Graphic* 54 (1896): 461.

an untapped energy, a "little dynamo."[43] In contrast, in an appended conclusion titled "A Lesson for Mothers," he recounted the story of a German mother who took her baby home from the incubator ahead of schedule, only to have it wither away so rapidly as to force a hasty return. That infant, Brisbane asserted, had suffered not from mere irregularity of temperature but from "brain fatigue" occasioned by excessive handling and activity. "Mothers would do well," he admonished, "to remember that the chief thing in caring for a baby is to keep its brain quiet."[44] For Brisbane, the incubator taught that excessive maternal stimulation and nurture could be a liability.

Yet the sideshow dimension of these exhibits remained a troubling undercurrent beneath their continuing success. Couney's decision to set up a permanent show at Coney Island's Luna Park in 1903 carried the process of popularization a step further. His rationale was to establish a convenient base from which to conduct his traveling shows. Luna Park provided, however, a decidedly different context from the international exhibitions. Directed at the working classes, its audacious colors, thrill rides, and sideshows represented what historian John Kasson has called a cultural revolt against genteel standards of taste and conduct.[45] Technology existed not to educate but to excite the senses. The midgets, giants, and freaks populating the resort bestowed it with an exaggerated, almost surreal character. One of the most popular shows at Dreamland, another Coney Island park where Couney opened a second incubator show, was Lilliputia, a miniature village scaled down to the size of its 300 human dwarf "inhabitants."[46] The popularity of Couney's own shows in such a setting suggests that they afforded more than education. Couney's Luna Park exhibition, in fact, became the longest-running show at Coney Island; it lasted well past the park's heyday and into the 1940s, the years in which premature infant nurseries finally were becoming a national priority.[47]

Although later in his career Couney insisted that he had done no more than make "propaganda for the proper care of preemies," the message he intended was not necessarily the message perceived. He claimed that obstetricians always referred him more babies than he could ever handle and that his patients had no other viable options for incubator care.[48] Without calling his sincerity into question, it is worth noting that the latter point could be made regarding the other inhabitants of "freak shows." Fairs afforded such persons an option for employment of last resort, one often demeaning but sometimes (especially in the case of "celebrity" midgets dressed up as European aristocracy) attended with a strange mixture of pretense and honor. Again, most accounts agree that Couney consistently tried to minimize sensationalism, reprimanding his barker-lecturers when they became too flamboyant. Yet the "spiels" that have been preserved did not completely eliminate any place for the public's fascination with oddities of nature:

> You may talk, ladies and gentlemen, you may cough. They will not hear you. They do not even know you are here. . . . Now this little baby came in nine days ago. It weighed only one pound eleven ounces and we were afraid we might be too late. It was even bluer than that little fellow over there in the other incubator. . . . Yes ma'am, it was a premature birth. . . . a little over six months.[49]

Much of this testimony was recorded in Couney's later years, by which time he was attaining a certain amount of respect among medical circles. Less is known regarding the extent to which his early exhibits provoked any public outcry. At least one group of child advocates, however, were not amused.

The New York Society for the Prevention of Cruelty to Children (SPCC) did, in fact, seek to outlaw Couney's shows. No sooner had the incubator doctor arrived in Coney Island than the society's Brooklyn chapter inquired into his license and the nature of his activities.[50] In 1906, the state organization attempted unsuccessfully to pass an amendment to the New York Penal Code "prohibiting the exhibition of infants undergoing the process of artificial incubation."[51] Its president, lawyer John D. Lindsay, struck again when a 1911 fire in Dreamland destroyed the park and forced the emergency evacuation of its incubator babies. The plaster-and-hemp-fiber construction em-

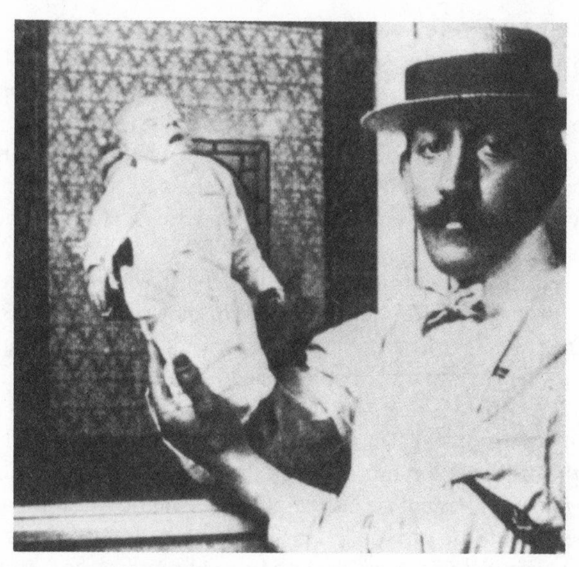

Premature baby on display, Buffalo Pan-American Exposition, 1901. *Source: Pediatrics* 64 (1979): 133.

ployed in its buildings made fire a regular visitor to Coney Island. The 1911 blaze ironically began in the Hell's Gate attraction, next to the incubator building, and created such panic and confusion that mistaken rumors of six infant deaths made the headlines of the *New York Times* the following morning.[52] Lindsay, a man of impeccable social credentials who wrote articles on colonial history for recreation, hardly sympathized with the culture of pleasure represented by Coney Island; he served, moreover, as legislative advocate for societies protesting cruelty against both animals and children.[53] In an angry letter to the *Times* just after the fire, he charged that the infants owed their survival to pure chance. Lindsay recounted the SPCC's history of opposition to Couney's incubator show:

> The society's investigation convinced me that the motive for the exhibit, which was presented as "charitable," was purely mercenary, violating every principle of medical or professional ethics; it was a "side show," advertised by the proprietors of the resort as one of the attractions with its "Shoot the Chutes," "Razzle-Dazzle," and "Monkey Theatre."[54]

Nothing apparently issued out of his protest, since Couney's exhibit at Luna Park survived. But it did suggest how his enterprise was perceived in polite circles.

The lesson thrust upon Couney was that as hard as he might try to deny his social context, he could not escape it. Just as the promoters of the Lion incubator claimed that it could transcend many of the limitations of its environment or caregivers, Couney may well have thought he could transmit an "educational" message on the Midway. But the setting that housed the incubator changed the message it sent. The public would inevitably perceive an incubator on the Midway very differently from one in a hall of technology. And Couney seemed to a remarkable degree oblivious to this reality. A similar insularity shielded him from seeing the tendency of his enterprise to neglect the mother in his quest to save the infant. He complained, in fact, that some mothers were all too happy to postpone the day of discharge, to the point that on a number of occasions he had to drive the babies home himself. Couney boasted in his 1939 interview that he had yet to be stuck with one.[55] Whatever his circumstantial associations with Budin, he was certainly far in spirit from the French obstetric tradition.

How American physicians would react to his work, and to the idea of incubator shows in general, was a more complicated matter. It hinged not only on Couney but on his most ambitious rivals, the promoters of the most elaborate show of all, the 1904 Louisiana Purchase Exposition.

American Physicians and Incubator Shows

The reaction of the medical profession to the incubator shows was at first positive. On at least two occasions the shows provided pediatricians unusual opportunities to employ the sophisticated equipment and environment of the Lion incubator station. The Buffalo Children's Hospital operated a modest incubator station for several years, using equipment purchased from the fair.[56] Its pediatrician, DeWitt Sherman, incorporated not only the Lion incubator but a number of techniques apparently learned from observing the fair. He relied on breast milk from wet nurses and asserted that the Lion's use of pure outside air saved his infants from a hospital measles epidemic. Sherman's debt to Couney was most apparent in his advocacy of nasal feeding, a technique rarely practiced outside of the fairs. He wrote little beyond a single article on his nursery, however, which treated a total of only twenty-nine infants between 1901 and 1905.[57]

The case of pediatrician John Zahorsky at the Louisiana Purchase Exposition provided the most striking example before the First World War of an American physician collaborating with an incubator show. The son of Hungarian immigrants who had moved to Missouri early in his life, Zahorsky became involved in premature infant care through his collaboration with Bayliss's exhibit at the St. Louis exposition of 1904. At the time, he was a general practitioner well on his way to limiting his practice to pediatrics, and within another year he would begin treating children exclusively, as clinical professor of pediatrics at Washington University.[58] Zahorsky was summoned to manage the show in the wake of a devastating epidemic of summer diarrhea that struck in July. The resulting mortality rate of nearly 50 percent produced a scandal in the popular press, as a small army of self-appointed incubator show "specialists" arrived to berate the exhibit's management. Zahorsky identified these individuals as showmen who had conducted incubator exhibits in other fairs but had failed to win a contract at St. Louis; their existence suggests that such shows were more widespread in the early 1900s than the few major expositions that attracted attention from the national press. At any rate, the exhibit's directing physician and promoters spared no expense in assailing the outbreak, adding to their already generous technological armamentarium a glass partition separating the visitors and infants. The mortality rate began to decline in September, but not before the controversy forced the resignation of the medical director. It was at this point that Zahorsky was recruited to take over.[59]

Zahorsky brought the perspective of an aspiring academic physician to the context of the incubator show. He used the experience to produce a series of articles that he later compiled into a monograph, *Baby Incubators: A Clinical Study of the Premature Infant,* published in 1905.[60] One could argue that this book, replete with references from the United States and especially Europe, represented the most thorough American synthesis on premature infant care to appear before Julius Hess's 1922 textbook, *Premature and Congenitally Diseased Infants.*[61]

Zahorsky went to great pains in the work to distance himself from the entertainment aspects of the fairs. He acknowledged that the great majority of his fellow professionals were opposed to the idea of "making an exhibition of human misfortunes, especially in the shape of tiny infants."[62] Zahorsky nonetheless offered three reasons why incubator institutes could still be beneficial. One echoed the argument made by Couney that the exhibits fulfilled an educational function. Zahorsky had no doubt that they had played a major role in calling attention to the problems of prematurity in the United States. The second rationale had to do with economics. "Money is necessary to save the lives of premature infants," Zahorsky contended; "It is questionable whether any hospital or asylum could undertake this work unless the State would give liberal support." The level of care provided by the show was expensive by contemporary standards, amounting to fifteen dollars a day. Zahorsky argued that such care represented a "boon to the poorer class." It also could be a boon to the professional class, for his third justification was that the incubator show could become a place to advance science.[63]

It was by linking the incubator show to a program of clinical research that

Incubator show supervised by John Zahorsky, St. Louis Louisiana Purchase Exposition, 1904. *Source:* John Zahorsky, *Baby Incubators: A Clinical Study of the Premature Infant, with Especial Reference to Incubator Institutions Conducted for Show Purposes* (St. Louis: Courier of Medicine Press, 1905), 19.

Zahorsky moved beyond Couney. He found himself in possession of the largest incubator station in the country. The exhibit employed a dozen Lion incubators, fourteen nurses working in three shifts, a resident physician, and a twenty-four-hour ambulance service with a portable incubator. Anxious to avoid scandal, the businessmen who promoted the show were willing, in Zahorsky's words, "to do anything that science has taught was necessary." Indeed, from his perspective they went beyond what was necessary, purchasing expensive gadgets such as a gas pressure pump and electric water heater.[64] He nonetheless had access to a remarkably well-equipped and controlled environment, a kind of laboratory of neonatal medicine thrust into the Midway.

Zahorsky took advantage of this setting to conduct studies that were more sophisticated than those of virtually any of his American contemporaries. Most of the articles on premature infants in the United States offered recommendations on care rather than addressing specific research questions. Zahorsky was unusual in calling attention to the inconsistencies between these reports, as well as those of Europeans. For example, he articulated the unresolved questions surrounding the incubator's temperature more forcibly than anyone else in the United States before the First World War.

Various authorities recommended keeping the incubator anywhere between 77° and 95°F and seemed to rely on clinical assessment as much as rectal temperature in setting the device. Zahorsky tried to develop recommendations based on infant body weight that accepted the maintenance of a rectal temperature between 97° and 99°F as a standard. Judging from the published literature, he was virtually alone in formally comparing individual temperature curves to answer the question.[65]

Other examples of his research illustrated Zahorsky's desire to interact with academic pediatricians. Particularly advanced from the context of his time was his use of growth curves to study nutrition to establish the caloric requirements of premature infants.[66] In this case, he was engaging a small but growing research tradition that emerged in Boston in the wake of Thomas Rotch's work on artificial feeding. Zahorsky identified himself with a newer school of thought in pediatrics that shifted attention away from the composition of milk to its caloric content; again, his writing reveals him to have been in touch with the current concerns of the academic leadership.[67]

But instead of becoming a father of neonatology, Zahorsky lived to see his work forgotten. In spite of later publishing three other books and over 100 articles and becoming famous for describing the two pediatric infections roseola and herpangina, he always considered *Baby Incubators* to have been his most important scientific contribution. His contemporaries disagreed, for Zahorsky found that "pediatrics at large rather frowned on this work."[68] Much of the problem doubtless had to do with its regional origins. Some local practitioners were proud of his research, remarking that St. Louis was about to become a new center of incubator care.[69] Outside of Missouri, the reception was more muted. His articles were published in a local journal, and his privately published book was ignored by most mainstream pediatric journals.[70] The principal organization of academic pediatrics, the American Pediatric Society, was an elite society whose membership was limited almost entirely to cities on the East Coast, especially New York, Boston, and Philadelphia. The society had already rejected Zahorsky's application for membership in 1900, after he had collaborated with a poorly regarded St. Louis physician on a paper addressing oral diphtheria antitoxin. He even had to present his famous paper describing the rash of roseola through an intermediary.[71]

Moreover, the antagonism between Zahorsky and the American Pediatric Society was mutual. In accounting for the poor reception to his work, it must be added that his own rhetoric was hardly calculated to appeal to the pediatric leadership. Specifically, he was hostile to one of his profession's foremost reform goals, that of making pediatrics a hospital-based specialty. For Zahorsky believed that hospitals were dangerous to infant life. He never became an advocate of hospital incubator stations and always believed in the superiority of home to institutional care for premature infants wherever possible.[72] Indeed, the St. Louis pediatrician was considerably more sympathetic than many of his professional colleagues to the capabilities of the mother to care for a sick infant. With time he became an aggressive proponent of "mothering" and challenged the belief of the pediatric mainstream that mothers could spoil their infants through rocking them in a cradle.[73]

Whatever the source of his estrangement, Zahorsky's defense of the incubator show included an unflattering comparison with the infant hospital, the setting where many of the most influential pediatric leaders worked. For Zahorsky, the real objection to incubator shows was not professional squeamishness but the inherent danger of institutionalizing infants. He attributed the diarrhea epidemic that had struck during the summer to "hospitalism," thereby linking it to a well-known phrase connoting the high mortality rates so common in infant hospitals. While conceding that meticulous aseptic technique and elaborate equipment could prevent such mortality, Zahorsky ultimately preferred the home as safer and far less expensive.[74] He pointedly added that though the businessmen funding the St. Louis exhibit knew nothing about the care of premature infants, "neither do the capitalists who build our hospitals possess any special knowledge concerning care of the sick."[75] The incubator show might involve a kind of pact with the devil, Zahorsky implied, but no more so than working in an infant hospital.

Zahorsky nonetheless personally believed that *Baby Incubators* failed mainly because of increasing professional disdain for incubator shows. The irony was that the Louisiana Purchase Exposition played a major role in this turnaround of opinion. It had begun as the most elaborate and "scientific" of all, but after the epidemic the exposition became a symbol of infant exploitation. The show's financial backers ended up losing money as well as facing embarrassment, discouraging would-be imitators.[76] It may have been joined by other examples that so far have eluded the historical record. In 1911, John Lindsay, of the New York SPCC, seemed to have St. Louis as well as other examples in mind in explaining why the incubator shows had failed outside of Coney Island. Though he noted that attempts had been made to create such shows in various areas of the country, they had failed, "owing to the popular indignation which resulted from the announcement."[77] The wave of early enthusiasm for incubator shows was clearly receding by the second decade of the new century.

After 1910, Martin Couney increasingly played the role of exile rather than pioneer, a curious anachronism in a nation whose professional leaders were increasingly turning against the Lion incubator. The new consensus took shape within the rapidly evolving specialties of obstetrics and pediatrics. Couney's incubator shows did little more than provide an environment where

Lion's device continued to survive underground during these years. Aside from the 1915 San Francisco exhibition, he found no further opportunities after 1910 to set up shows in major world fairs in the United States until 1933. Instead, he was forced to set up exhibitions in local amusement parks in locales such as Atlantic City, Denver, and Chicago.[78] Coney Island itself was losing some of its distinctiveness by the 1920s as new forms of entertainment, particularly movies and radio, emerged to satisfy the needs of popular culture.[79] Couney maintained a low profile during these years, never publishing any account of his work.[80] Most of the medical profession ignored him. Those who accepted his successes attributed them to the availability of breast milk rather than incubators.[81] If anything, the existence of incubators in fairs made them suspect as a form of medical quackery. One pediatrician announced in 1917 that "incubators are passé, except at country fairs and sideshows."[82] To many American physicians, the elaborate "scientific" appearance of the incubator doubtless explained its persisting attraction to the American public, much as static electricity machines had mystified and excited the previous generation.[83]

The one exception to this overall pattern of decline in Couney's fortunes was his meeting and growing friendship with Julius Hess of Chicago, the man who became the leading figure of American neonatal medicine in the 1920s. Though accounts of their first meeting are conflicting, most likely it took place when Couney opened an incubator show at Chicago's White City amusement park in 1914. According to one account, the Chicago Medical Society refused to allow Couney to set up his show unless supervised by a local pediatrician. Hess was chosen for the job.[84] Unlike Zahorsky, the Chicago pediatrician had already developed an interest in premature infants. He had operated incubators at Michael Reese Hospital since at least 1907, published his first article on the feeding and care of premature infants in 1911, and created his first incubator (a model that owed more to Credé and Rotch than to Lion and Couney) in 1914.[85] Over the course of the years, the two men developed a close friendship, which Hess eventually openly acknowledged. He was slow to do so at first, however. In the introduction to his 1922 textbook, Hess's only explicit mention of Couney was an inconspicuous addendum to his introduction, in which he noted, "I desire to acknowledge my indebtedness to Dr. Martin Couney for his many helpful suggestions in the preparation of the material for this book."[86]

In retrospect, Couney represented one of a number of streams that converged to create Hess's incubator station. The two men particularly came to identify with each other in the popularization of premature infant care. The two careers truly merged in 1933, when they collaborated to set up an incubator show at the Chicago Century of Progress Exposition that went far to promote the fame of both men. At that point, the Chicago pediatrician's support was critical for the rehabilitation of Couney's reputation in professional opinion. Couney would thus undergo a final transformation in the 1930s to the role of the forgotten and lone pioneer of neonatology. On the occasion of his last great show at the 1939 New York World's Fair, the city's medical profession and Julius Hess honored him in a special banquet, and the *New Yorker* made his story the subject of an entire feature article. Soon thereafter, he closed his Coney Island show, claiming that improving hospital care for premature infants had finally rendered it unnecessary.[87]

❧

The incubator show movement may have been the most spectacular means by which the device crossed the Atlantic, but its significance as a mode of technological transfer proved to be no more lasting than the fairs themselves. It would surely be inappropriate to reduce either Martin Couney or John Zahorsky to medical "missing links" between Pierre Budin and Julius Hess. But if the story of the fairs turns out not to be very important in terms of direct influence, it still remains worth understanding as a parable. The image from this strange interlude that continues to arrest our attention is that of John Zahorsky, the aspiring academic and clinical investigator, studying premature infants on the Midway. His was certainly an unusual career, one that bore little in common with the academic leaders who dominate the remainder of this history. At the same time, there remains a dimension in which his experience captures the way in which the premature infant program was changing in its passage from France to the United States. Whereas French obstetricians had promoted the incubator as part of a public health program to reduce premature infant mortality, American physicians viewed it primarily as a symbol of science.

This enthusiasm for science needs to be understood in the context of the American medical profession during the Progressive reform period. To a far greater extent than had ever been the case before, physicians in the United States during the early twentieth century accepted the ideal of medicine as an applied science. The therapeutic breakthroughs of the 1890s, particularly diphtheria antitoxin, no doubt explain much of the widespread excitement over the potential for science to transform medical practice. But the appeal of science went beyond its demonstrated practicality. It offered a strategy to elevate dramatically the profession's prestige and transcend its divisions, to replace its trade-school educational system with formal academic training. The laboratory, reformers believed, would train the physician to think like a scientist; ethical codes would become obsolete as scientific training provided the values of precision and honesty necessary to guarantee professional integrity.[88] These ideals swept the country in the wake of the medical school reforms pioneered first by Harvard in the 1870s and then (with more fanfare) by Johns Hopkins in 1893, climaxing in the publication of the Flexner Report on medical education in 1911. Abraham Flexner finalized a process that had long been gaining momentum, the incorporation of the sciences into the medical curriculum.[89] Science promised American physicians a collective degree of prestige and income they had never enjoyed in the past, one which public health could not rival.

Technology played a critical symbolic role in the rise of scientific medicine. It did so because the profession's virtual unanimity on the virtues of science was not matched by agreement over what science meant. In medicine, as in other professional disciplines during the Progressive Era, science could imply a variety of methodologies, ranging from observation to statistical analysis to laboratory experimentation.[90] Yet the ideals of objectivity and precision united many of these specific approaches. As Stanley Reiser has argued, diagnostic instruments were of fundamental importance in transforming the nature of medical practice in accordance with these values. New devices such as the electrocardiogram that translated physiological processes into numeric and graphic forms seemed to make them more permanent, more objective, more independent of the observer's limitations.[91] The Lion incubator's defenders invoked a similar kind of reasoning for therapeutic technology. They regarded the device as scientific to the extent it freed the infant from the limitations not of the observer but of the caretaker. It promised to create an environment that theoretically conformed to the principles of science rather than the idiosyncracies of the home and family.

The reverse side of this powerful scientific idealism was that it could encourage the well-intentioned physician-investigator to deny any dependence on social context. It beckoned physicians to push the incubator to the boundaries of what was technologically possible, to create a device that could transcend the limitations of its environment and caretakers. The excitement could be so intoxicating that it led an undoubtedly sincere clinician such as Zahorsky to rationalize away whatever ethical qualms he might have had over displaying premature babies in an amusement park. His story in the end had a tragic as well as a heroic aspect. It also had some intriguing parallels with that of the other leading figure in early-twentieth-century American neonatal medicine, the obstetrician Joseph B. DeLee. For DeLee also saw himself as an isolated crusader for science rather than a pragmatist ready to accomodate to social reality. The story of his remarkable achievements must be set against his failure to introduce the French program of obstetric neonatology to the United States.

CHAPTER SIX

The Experiment in Obstetric Neonatology

In France, obstetricians dominated the discourse concerning the incubator well into the early 1900s. More precisely, France gave rise to an obstetric profession that considered the care of the newborn a high priority. Stéphane Tarnier, Pierre Budin, and many of their allies were in a sense perinatologists, physicians who cared for both the mother and the infant. As depopulation anxiety rose to a crescendo at the end of the century, the balance of priorities shifted. Budin spoke as if the conquest of puerperal fever had essentially eliminated the high mortality associated with childbirth; obstetricians, therefore, needed to move to the new frontier represented by the infant. Whatever consequences this outlook had for the mother, it certainly promoted interest in the newborn. For obstetricians did, in fact, enjoy critical advantages over pediatricians in caring for premature infants. Most notably, they worked in maternity hospitals where they enjoyed immediate access to the infant, as well as the potential for breast-feeding and maternal involvement. Maternity hospitals also singled out newborns as a special population distinct from other babies. As a result, French obstetricians acquired a reputation as European leaders in the care of premature infants.

One of the critical questions for understanding the evolution of incubator care in the United States is why the American obstetric profession chose a different path with respect to the newborn. Was it possible that American neonatology could have evolved as a domain of obstetrics rather than of pediatrics? During the early years following the introduction of incubator care, particularly between 1890 and 1905, many obstetricians would have answered in the affirmative. Their involvement deserves close examination from two standpoints. First, obstetricians developed a relatively aggressive style of managing the premature infant that might be described as mechanistic. Though it won few adherents in the short run, this approach remained an alternative to the environmental philosophy that would become dominant as pediatricians took over management of the newborn. Second, in a negative sense, obstetric claims to the newborn clashed with those of pediatrics. The conflict between the two shaped how each perceived the premature infant and, ultimately, the incubator.

A Domestic Technology

The story of obstetric involvement with the incubator really begins with general practice, since obstetrics in the late nineteenth century represented a part-time practice more than the formal specialty it later became.[1] To be sure, the same point also applied to pediatrics in the same time period.[2] But it makes more sense from our perspective to concentrate especially on the continuity between general practice and obstetrics. With respect to childbirth, both types of practice shared a primary focus on the mother as opposed to the infant. In the United States, the physician delivering an infant traditionally handed it over to the mother or her female attendant once breathing had been established. By the late 1800s, middle-class mothers often hired older (and often widowed) women as "nurses," much as they might hire a servant.[3] But the physician's primary concern was the mother rather than the newborn.

No doubt, most physicians retained this focus on the mother's welfare and interests well after the introduction of the incubator. Given that standard obstetric textbooks throughout the 1880s contained almost no suggestions regarding the care of the premature infant, the average physician most likely continued to defer to the mother and nurse.[4] The comments of contemporary physicians suggest that many mothers recognized the importance of preserving the infant's body heat through simple means, such as rubbing the naked infant before a fireplace or employing a padded basket with hot water bottles. It is likely that such measures remained more prevalent than the incubator through the 1890s.[5] Nonetheless, a minority of obstetricians and general practitioners advocated the new technology while continuing to retain a primary emphasis on the mother. Their attitude can be characterized as "conservative obstetrics," to emphasize its continuity with the values of mid-nineteenth-century practice. The use of this term stresses the close link between medical conservatism, in the sense of opposing excessive intervention, and social conservatism, in terms of deferring to the mother rather than the infant; conservatives were more ready to sacrifice the infant than to endanger the mother. With regard to the infant, this ideology fostered a domestic style of technology promoting simple incubator designs that differed little in principle from a padded basket.

The clearest example of this orientation was to be found among those who linked the incubator to the earlier procedure of artificially induced premature labor. The third quarter of the nineteenth century, it will be recalled, generated an extraordinary degree of enthusiasm for this procedure in situations where a narrow pelvis might otherwise require craniotomy, forceps, or other invasive measures. Yet in the United States, its advocates rarely addressed the care of the resulting premature infant, the implication being that its management was a problem for the mother rather than the physician. In France during the same time period, Tarnier's interest in the premature infant originated as an offshoot of his maternity hospital reform efforts to a much greater extent than his advocacy of induced premature labor. Pierre Budin in 1887 contrasted his own and other French physicians' preference for premature induction with the German advocacy of cesarean section and yet did not emphasize the linkage of the two issues in his papers on the premature infant until the late 1890s.[6]

Yet although premature induction did not in itself motivate physicians to search for new methods of treating premature infants, the rise of incubator therapy certainly reinvigorated the older procedure. The obstetrician Henry Bettmann of Cincinnati pointed out in an 1891 address that the incubator made premature induction more advantageous than ever. Examining how both maternal and infant risks from induced premature labor had declined over the previous two decades, he concluded enthusiastically that "what asepsis has done for the one the various incubators have done for the other."[7] Many American obstetricians promoting the incubator also advocated premature induction, although few spelled out the connection as clearly as Bettmann.[8] The obstetrician who published the largest American case series examining the incubator before the First World War, James Voorhees, became best known among his professional colleagues as the inventor of a new balloon to induce premature labor. Voorhees published both his incubator study and his first paper on the "Voorhees bag" in 1900.[9] There was a sense that for these obstetricians the premature infant would not have been such a relevant problem had it not been connected to a procedure aimed at the mother.

Just as conservatives tended to rationalize the benefits of the incubator with regard to the welfare of the mother rather than that of the infant, they acknowledged her traditional role in rearing the infant after birth, as well. Significantly, American physicians who used the language of conservative obstetrics to promote the incubator often preferred the simple and technically unsophisticated Auvard model heated by hot water bottles. This design required such constant attention that it represented, as noted in France, an extension of the nurse's capabilities, rather than a replacement. In the middle-class home setting, it reinforced the centrality of the mother and her nurse to the welfare of her infant. W. Byford Ryan, an Indiana physician, expressly preferred simple homemade incubators without thermostats. Such devices were preferable "for the reason that the nurse would not be compelled to pay so strict attention to the infant were the apparatus automatic."[10] Ryan's emphasis on the infant's female caretakers went well beyond the supervision of the incubator. The mother's presence was essential in order to overcome the problem of keeping the infant warm during the critical hours after birth awaiting the construction of the device. Ryan insisted that, as a matter of "utmost importance," the infant be immediately placed in bed with its mother or another woman "who will devote attention to imparting heat to its body, by keeping it in close contact with the anterior portion of her own person." Later, the mother remained essential for the provision of nutrition (breast milk) and for meticulously watching over the precarious infant.[11]

To the extent that they concentrated on the mother, American obstetricians also evaded some of the ethical questions related to the long-term consequences of saving premature babies. The task of assisting the premature infant born of an induced delivery represented a relatively straightforward moral enterprise, since the infant was clearly a product of the obstetrician's art rather than its own constitution. Many physicians believed that such infants did better than their "naturally born" counterparts who, they rationalized, suffered from a hidden cause analogous to syphilis that drained the infant of its life while still in the womb.[12] The emphasis on maternal love was also relevant in this regard; physicians were unlikely to have qualms about propagating poor heredity when dealing with the plight of a childless woman with traditional Protestant virtues. The prospect of saving a child in the meager circumstances of an urban tenement, where few resources were available and the infant's outlook uncertain, raised far more troubling questions. "We sacrifice the child," wrote one obstetrician on the analogous subject of craniotomy, "but then we know that almost all these cases are among the poorest classes," of whom "scarcely one-half live five years."[13] It was in this type of setting that many such infants were most likely, in the words of one physician, "quietly laid away" in an unobtrusive corner of the household.[14]

The Tarnier-Auvard incubator thus found a niche in the American medical landscape. It did so, however, principally in a setting identified with the forces of conservatism. And though a steady trickle of articles on home incubators continued to appear throughout the early twentieth century, they excited little attention among obstetric leaders. Obstetrics in the United States was evolving in a very different direction from that in France. It was identifying itself with a certain conception of "science," a factor of considerable importance for its response to the incubator.

The Rise of Professional Obstetrics

Though the practice of obstetrics generated a considerable proportion of the income of many nineteenth-century American physicians, it held only limited appeal as a formal specialty. For most physicians, full-time obstetrics suggested midwifery rather than medicine. Midwives had traditionally been responsible for most births in the United States and had practiced a relatively conservative style of management that generally relied on nature to guide the progression of labor. Yet since the beginning of the century, they had been losing ground to physicians in terms of popularity, to the extent that by 1900 physicians probably attended half of all births. Women tended to call upon physicians for two reasons: the promise of new interventions to make childbirth safer and more painless and the association of medicine with science. American midwives never succeeded in professionalizing, particularly with regard to developing formal standards for training and self-regulation. They also failed to incorporate the prestige of science into their training and practice, in pointed contrast to their counterparts in France. Any physician contemplating specialization in obstetrics had to find a strategy to distinguish his practice from the perceived second-class status of midwifery.[15]

In meeting this challenge, obstetric specialists saw surgery as providing the model for uplifting their specialty. For surgery, by the 1890s, had achieved an enviable degree of prestige for a profession that had long been relegated to the status of a craft. The rise of anesthesia and then asepsis had generated a dramatic expansion of operative procedures, ranging from appendectomy to mastectomy. By the 1880s, physicians sniped how illnesses such as "bellyache" had become surgical diseases. Surgeons accounted for an increasing proportion of hospital admissions and consequently gained increased power within that setting as well.[16] For would-be obstetricians, the moral behind these events was to develop their own specialty along similar lines. Obstetrics was to be combined with gynecology. The American Association of Obstetricians and Gynecologists was first organized in 1888, eleven years after the founding of the American Gynecological Society.[17] Though the relationship between obstetrics and gynecology remained contentious well into the twentieth century, their common surgical orientation was rarely at issue.[18]

Thus, American obstetricians, in choosing to ally with surgery rather than pediatrics, took a strategy quite different from that of their French counterparts. The contrast was admittedly not absolute; in France, as we have seen, Tarnier's rise to eminence was closely linked to his incorporation of surgical asepsis, as well as mechanical interventions such as his improved forceps. Yet it still represented a significant difference in emphasis. The American obstetric profession would never produce an individual with as serious an interest in newborns as Pierre Budin's. Two factors were especially important in explaining this outlook.

The first factor distinguishing the contexts of French and American obstetrics has already been mentioned: the simple fact that the preservation of infant life did not represent a national priority in the United States as it did in France. In a nation where immigration more than compensated for any decline in birthrate, depopulation was hardly an issue. Americans spoke instead of "race suicide," the prospect that the native-born, Anglo-Saxon population would be overwhelmed by foreigners.[19] Amplifying such nativist and racist fears was the persisting power of social Darwinist arguments that infant mortality campaigns would result in the survival of the "unfit." Obstetricians, already stigmatized by their association with midwifery, saw little to be gained by allying with pediatrics.

Second, American physicians were increasingly unlikely in the late nineteenth century to see French medicine as a role model for their own profession. One reason Tarnier's apparatus had trouble exciting the American medical elite was that French medicine was no longer synonymous with medical progress. By the 1880s, the German university had replaced the Paris hospital as the chief source of inspiration for the younger generation of physicians, and the laboratory, rather than the clinic, became the symbol of a new era of scientific medicine.[20] As a consequence, few American physicians had any direct experience with Tarnier's hospital service or his technology. Indeed, many actively expressed their disdain for French clinical medicine. One editor, in a vitriolic outburst, pronounced French medical journals to be the worst of any civilized country, their poor quality reflecting their great number.[21] Another likewise remonstrated that French medical periodicals represented "prolix and, for the most part, profitless reading, and exceed in number the legitimate demand."[22] One might justifiably make the same point with respect to the United States, which produced more journals and articles than did any other country in the world.[23]

Although differences in temperament doubtless helped to separate American physicians from their former mentors, the intensity of their rhetoric suggests that Americans to some extent projected the shortcomings of their own profession on the French. The divided and contentious world of Paris mirrored the political and geographic divisions splitting American medicine; French theoretical disputes reminded Americans of their own antebellum struggles between allopaths and sectarians. Paris medicine represented the banner for the conservative wing of the American medical profession as the younger wing rallied around Germany and laboratory medicine.[24]

The decision of the obstetric profession to pursue a marriage with surgery rather than pediatrics is hardly surprising in light of these factors. Nor do we need to belabor the point that to the extent that the surgical paradigm represented a distraction from the pediatric, it obviously discouraged obstetric interest in the infant. But surgery shaped the way obstetricians understood the incubator in a positive sense as well. Surgeons were naturally more predisposed to use tools and machines. Moreover, they were consciously reevaluating their procedures in the late nineteenth century as part of a movement toward what they called scientific obstetrics. As a result, new values emerged that would profoundly affect attitudes toward the incubator.

As obstetric leaders sought to justify the need for specialization, they often framed their differences from ordinary practitioners in rhetorical terms opposing scientific and traditional obstetrics. Although both conservatives and reformers condemned unnecessary interventionism, the "meddlesome midwifery" so rampant in the 1880s, they proposed different solutions. Conservatives appealed to the ideal of "watchful waiting" popular in the mid-century. The best strategy to avoid unnecessary intervention, they contended, lay in understanding and respecting the natural course of labor.[25] Reformers turned this logic on its head, arguing that the key point was not to withhold intervention but to conduct it under controlled circumstances. They emphasized that the worst time to intervene was late in the course of labor and that "scientific obstetrics" above all meant early and systematic intervention.[26] For example, the proponents of scientific obstetrics argued that the fearsome reputation of cesarean section stemmed from the fact that it was reserved only for desperate situations. It was far better, they insisted, to anticipate problems and introduce the procedure early to minimize mortality.[27] Such arguments mirrored those used by proponents of scientific surgery, such as William Halsted of Johns Hopkins, whose advocacy of radical over simple mastectomy can be seen in similar terms.[28]

Obstetric specialists also came to believe that the program of systematic intervention required moving childbirth to the hospital. The United States had previously lacked the large maternity hospitals that had given rise to the French premature infant campaign. Throughout the nineteenth century, only desperately poor women, often unmarried and without family support, sought "lying-in" hospitals as a refuge. Most maternities maintained a tenuous existence beyond the bounds of respectable society, operating in the cramped quarters provided by a converted home or tenement, ravaged by puerperal fever, and ostracized by the community. The board members, matrons, and nurses who directed their operation (many of whom were women) frequently viewed their charges with a powerful sense of maternal and moral responsibility, resisting the growing tendency of medicine toward acute care and medical intervention.[29] Yet obstetricians increasingly saw maternity hospital reform as critical to their own success. The lying-in hospital offered an independent base from which academic obstetricians hoped to establish their specialty on equal footing with those of medicine and surgery. More important, it furnished an environment in which the physician could aspire to control every aspect of labor and delivery, more analogous to a laboratory than a charity.[30]

Although obstetricians achieved their reform goals only gradually, by the early 1900s their efforts were beginning to bear fruit. The first visible sign of this transformation was the construction of a series of new lying-in hospitals in New York and other prominent eastern cities beginning in the late 1890s. The new maternities were larger than their predecessors, typically located in buildings of several stories rather than in converted homes. They varied considerably on the inside, often expanding in erratic fashion, reflecting the ebb and flow of funding availability. Nonetheless, by the First World War a number of leading lying-in hospitals had acquired sterile operating rooms, isolation wards, laboratories, and private rooms.[31] The latter were especially critical, since middle-class patients eventually brought the revenue and prestige that enabled obstetricians to consolidate their profession in the 1920s and finally merge the maternities with general hospitals and medical schools.[32] During the first decade of the twentieth century, however, the victory of professional obstetrics remained uncertain. A similar shadow hung over the question of what role technology would play in the maternity hospital.

The transformation of American obstetrics and the impending rise of hospitalized childbirth had great potential implications for the infant as well as for the mother. It promoted values of precision and early intervention that could be applied to the care of the newborn. An obstetrician at a maternity hospital in San Francisco charged that incubators often failed at home because general practitioners waited too long before going to the trouble of finding one.[33] Another in St. Louis derided his conservative colleagues who cited "the rule that the best mechanic needs the fewest tools, and the best doctors the fewest drugs" to dismiss the utility of the incubator.[34] But the obstetrician who went to the greatest lengths to apply the paradigm of scientific obstetrics to the newborn deserves particular consideration.

Technology as a System: The Incubator Station

The most ambitious effort to link the incubator to this broader program of scientific obstetrics in the early twentieth century was that of obstetrician Joseph B. DeLee of Chicago. His story is worth recounting not because it was typical but because it illustrates the technological paradigm of neonatal care in its purest form. One of the most renowned and controversial figures in American obstetrics, DeLee achieved particular notoriety among historians of childbirth for his advocacy of the "prophylactic forceps" operation in 1921. Comparing the high mortality of women in childbirth to that of the salmon in spawning, DeLee inverted the metaphor of natural healing and argued that birth was a pathological process. He therefore proposed that it be managed through a series of systematic interventions under anesthesia, including induced labor, forceps delivery, and routine episiotomy.[35] The prophylactic forceps operation epitomized the early-twentieth-century ideal that sought to render childbirth a predictable event through standardized procedures. Indeed, if anything, it was too ideologically rigid for the many contemporary obstetricians who remained anxious to chart a middle course between the perceived extremes of meddlesome midwifery and watchful waiting.[36] DeLee's early-twentieth-century incubator station similarly embodied an ideal of neonatal care that made it an anomaly in its time and yet illustrated a paradigm of technological medicine that would triumph in a later day.

DeLee's Chicago Lying-in Hospital exemplified the volatile mixture of scientific idealism and zealous humanitarianism that drove his professional career.[37] The son of Jewish immigrants from Poland, DeLee spent his early years in the tenements of Manhattan's East Side before his family moved to Chicago. Throughout his career, he played the role of the outsider whose sense of identity required that he not merely imitate but exceed the standards of his professional colleagues. In medical school, he had a reputation for seriousness, hard work, and a sense of destiny; one acquaintance described the earnest Jewish physician as having a "Jesus complex."[38] DeLee was continually aiming high in his aspirations and compromising only uneasily with reality. After visiting a number of European maternity hospitals during his postgraduate studies in the 1890s, he returned to pursue with remarkable singlemindedness his greatest dream: the creation of a scientific maternity hospital for the Chicago poor. He translated his vision into reality only fitfully, dividing his limited time between fund-raising and a busy obstetric practice. Many of his professional colleagues showed little interest, while help came from less conventional quarters ranging from the settlement worker Jane Addams at Hull House to, most importantly, the Chicago Jewish community. In 1899, DeLee finally opened the Chicago Lying-in Hospital in a private residence on Ashland Boulevard, a modest building that he hoped would furnish a stepping stone to a far larger hospital closer to European standards.[39]

His incubator station opened soon thereafter and promptly was highlighted yearly in the hospital's annual reports.[40] "This is the only public incubator station in the country," he wrote in 1903. "In fact, it is the only attempt to make a 'Service des Débiles,' similar to those of Paris, in this country."[41] This statement was only partly accurate, for even at its peak DeLee's nursery never had more than four incubators. And although he may well have seen the *pavillon des débiles* at the Maternité during a postgraduate visit to Paris in 1894, he went well beyond imitation.[42] His incubator station embodied the scientific idealism of the world's fairs rather than the practical technology of its French counterparts.

The distinctiveness of DeLee's incubator station began with its hardware. Pointedly dismissing the Tarnier *couveuse* as little more than a shoebox with hot water bottles, DeLee developed his own version of the Lion incubator. He sought a device that would regulate heat automatically and at the same time provide a plentiful supply of "fresh, uncontaminated air" to protect the infant from infection.[43] Unfortunately, manufactured versions of the Lion fell far short of his expectations. DeLee investigated several available models of the Lion, such as that manufactured by the New York instrument maker Kny-Scheerer, and found their thermostats to be unreliable. Here, his own considerable technical expertise proved invaluable. A tinkerer since childhood, DeLee studied a variety of poultry incubators and developed an expanding capsule of ethyl chloride as his own thermostat.[44]

This thermostat marked only the first of many obstacles DeLee would encounter. The gas supply of Chicago Lying-in was no more constant than it had been at the Maternité. Whereas this problem encouraged Budin to give up the Lion model, it spurred DeLee to introduce a gas regulator. Ventilation complicated the maintenance of a constant temperature still further; in order to obtain outside air, DeLee installed a system of intake and exhaust pipes reminiscent of those used by Lion and Couney. During the Chicago winter, the temperature of the fresh air from "the sunny outside," which DeLee so ardently desired, could drop to 12°F below zero. He had to install a series of dampers to allow the nurse to compensate for sudden fluctuations of wind and temperature. Finally, he had to maintain the station itself at a fairly stable temperature in order to prevent radiative heat loss from the incubators.[45]

Yet DeLee did succeed in solving these various technical problems, and his small incubator station became the most advanced of any American hospital during the first decade of the twentieth century. His success illustrated that the technical problems of the thermostat and ventilation may have been too formidable for most physicians but were not beyond the capabilities of the time. He took pride in his accomplishment, attempting to patent his incubator and publishing an account in several medical journals.[46] "The claim is made for this apparatus," he asserted with pride, "that it is an incubator built on scientific lines, and not a mere warm box."[47] The device embodied for DeLee the ideals of systematic and rationalized intervention that marked the new scientific obstetrics. As with childbirth, the scientific treatment of the newborn could best be carried out in the controlled environment of the hospital. The fact that DeLee explicitly envisioned his incubator station as a system suggests a theme that resonated with the broader context of contemporary American technology.

Historian of technology Thomas P. Hughes contends that the dominant theme behind American technology since the mid-nineteenth century has been the construction of technological systems.[48] His metaphor works particularly well for electricity, a technology that clearly consists of components that must be integrated into a network serving a common goal. Hughes uses this concept to argue that Edison's invention of the light bulb was less important than the electric lighting system that sustained it. There were, after all, some twenty types of incandescent lamps invented between 1809 and 1878. Edison succeeded where others failed, Hughes argues, because he conceived of the problem holistically, in terms of solving a system of interrelated problems. The light was but one component of this system, arguably no more important than the generator, the main, and the feeder and the parallel distribution system that Edison developed as well.[49]

Hughes' system metaphor was not developed for medical technology, and one would not want to push too hard its relevance to DeLee's incubator system. But one aspect of his model offers a particularly useful insight regarding how DeLee developed his station. Hughes speaks of how system builders at a given point confront critical problems that hold back further development of the entire system. He describes these by the military metaphor of "reverse salients," referring to how offensive forces in warfare may fail to overpower a critical segment of the defensive lines.[50] What distinguished DeLee from many of the other physicians in this study was a constant drive to advance not so much scientific knowledge but its application. He sought to identify the critical problems, medical and logistical, obstructing his goal of treating the premature newborn.

DeLee first concentrated on the reduction of mortality in the first hours of life prior to hospitalization. Equipped with a battery of incubators and a staff of trained nurses, he turned to confront the perennial problem of the "outborn" premature infant. Like the *services des débiles* of the Paris maternity hospitals, DeLee directed his station not at the relatively few infants born in the hospital but at the far larger number born at home. The consequences were, perhaps, predictable. Of his first series of twenty-eight babies, eight died soon after having arrived nearly frozen or in convulsions.[51] The high percentage of early deaths inflated his overall mortality rates and embarrassed his efforts to obtain further support. He followed in this regard the example of his French predecessors by dropping from his statistics the infants who appeared doomed upon arrival. In his 1903 hospital annual report, sent to current and potential contributors, he therefore claimed that his survival rate was on the order of 90 percent, when it was really closer to 50 percent.[52] Yet it was not DeLee's nature to accept such high early mortality as an inevitable fact of life. Whereas Budin responded to similar setbacks by shifting his focus to larger, better-developed premature infants born in maternity hospitals, DeLee attempted to extend his technological approach to reach out into the home. He envisioned the first hours of life as a critical period

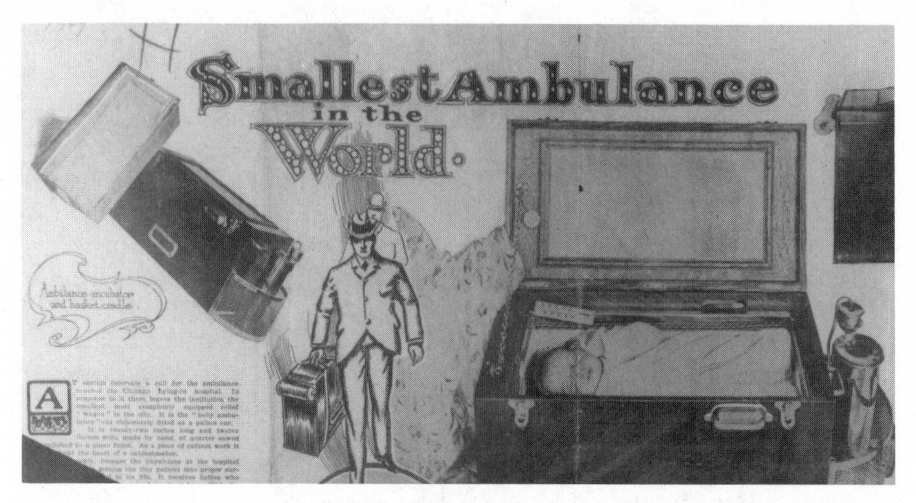

Joseph DeLee's transport incubator. *Source:* Northwestern Memorial Hospital Archives.

where failure could undo all efforts at later care. It represented, in a meaningful sense, a reverse salient.

The technological answer to the problem of home birth was DeLee's transport incubator.[53] In truth, this device represented little more than a Tarnier-Auvard incubator equipped with a handle. The strategy of seeing early intervention as the key to reducing mortality accorded well with the dictums of scientific obstetrics. DeLee's portable "ambulance incubator" could be dispatched with a physician and nurse anywhere in the city. Constructed of oak polished to a piano finish instead of the shining metal of the Lion incubator, it embodied the aesthetic values of domestic rather than scientific technology; one newspaper feature noted that its crafted appearance "would delight the heart of a cabinetmaker."[54] Its reassuringly conventional construction symbolized its position bridging the gap between the home and the hospital. DeLee looked forward to the day when every major city would have an incubator station (preferably connected with a lying-in hospital) to which "children could be brought, even from great distances, for that particular care which special training and practice only are able to bestow."[55]

Unfortunately, DeLee's clever solution to the transport problem only brought him to a second barrier, the treatment of respiratory distress. After the creation of the ambulance service, the station reached the apogee of its success by 1906. It admitted around twenty patients each year, of whom the number of those arriving "moribund" and dying soon thereafter decreased from a peak of eleven annually to less than three. Yet while the number of hopeless-upon-arrival cases declined, not only did the overall deathrate remain high, but roughly half of all these deaths continued to occur in the first thirty hours.[56]

Far more than most of his contemporaries, DeLee perceived that a physiological factor other than hypothermia caused many of these deaths. He proved himself to be an astute observer of the problem of respiratory distress suffered by the smaller premature infants on his service. Such infants turned a "reddish blue," grunting or moaning with each inspiration, forcing their chests to sink in at the base with each breath. He described them as atelectatic, the term used for newborns who failed to expand their lungs after a difficult resuscitation.[57] Congenital atelectasis, it will be recalled, was understood as a complication of asphyxia, rather than prematurity, in the nineteenth century. The treatment of asphyxia had long been a concern of obstetricians, who drew from a venerable tradition of mechanical interventions. The subject particularly fascinated DeLee, who was highly skilled at such difficult maneuvers as inserting a catheter into the trachea of an asphyxiated newborn by the sense of touch alone.[58] DeLee brought these skills, and the mechanistic orientation behind them, to the care of the premature infant.

DeLee's conceptualization of the premature infant in terms of the model of asphyxia rather than feebleness led to a remarkably interventionist style of treatment that contasted sharply with the dictums of conservative medicine. With an increasing sense of frustration, he applied all manner of resuscitation techniques to premature infants suffering from respiratory distress, including

> clearing the air passages; hot bath with frictions; artificial respiration by all the methods, including Schultze swingings; mouth to mouth insufflation, with a catheter in the trachea, first gentle, then forcible; the same with pure oxygen; continuous administration of pure oxygen (in one case 300 gallons were used); . . . incubator at high, medium, and low temperatures; dilatation of the sphincter ani; Laborde's tongue tractions; electricity to provoke crying; . . . allowing the cord to bleed; salt solution hypodermically; colonic flushings; administration of whisky, digitalis, etc., etc., and finally, no treatment at all.[59]

All of these methods were respectable alternatives to revive an asphyxiated infant. Many were covered in an article that DeLee had written in 1897, entitled "Asphyxia Neonatorum—Causation and Treatment."[60] Their reappearance here returns us to the nineteenth-century theme of the obstetric perspective of the premature infant. They also demonstrate the extent to which DeLee focused on the acute and early mortality of premature infants.

The flip side of DeLee's fascination with acute intervention was a corresponding lack of interest in the chronic issues accompanying prematurity. Premature infants, particularly those of early gestations, required nearly constant attention over a sustained time period. DeLee was hardly in any position to provide such care himself, nor did he possess resident physicians to do so. His solution to the problem recalled the traditional division of labor between physicians and women following the treatment of asphyxia, for it entailed once again handing the baby to the nurse. Yet in this case, turning over the infant to the nurse did not imply a limit to medical responsibility.

Far from being an anomaly within DeLee's incubator system, the nurse became one of its most critical components. Instead of pitting the machine and the nurse against one another, suggesting that the one might replace the other, DeLee integrated them. In the process, he sought to replace the nurse's individualized (but incommunicable) judgment with the predictability of standardized protocols. His essential ally in this project was the institution's superintendent of nurses, Emma E. Koch. Having arrived at the hospital in 1899 intent on starting a school for nurses, Koch was responsible for many of the practical issues attending the incubator station.[61]

Koch's collaboration with DeLee led to the publication in 1904 of the popular textbook *Obstetrics for Nurses.* Written when the incubator station was approaching its zenith, the book contained a chapter on premature infants that concentrated not on the construction of incubators but on the care of their occupants. It represented the most extensive American attempt of its time to set down standards of nursing practice for premature infant care. Whenever possible, it sought to replace the vagaries of individual judgment with objective criteria. Rather than simply affirming that feeding be tailored to the needs of the infant, for example, DeLee and Koch provided a table specifying exactly how much to feed an infant of a given age and size.[62] These guidelines did not completely exclude any role for feminine intuition. In a nursing journal article, Koch used this to her advantage, reminding her readers not to forget the importance of "womanly instinct in gently and tenderly handling the tiny bit of humanity."[63] But the main thrust of her writing was to shift emphasis from compassion to standardized care and scientific efficiency.

Koch in this regard reflected the growing interest of many of her middle-class professional contemporaries in redefining nursing in terms of a scientific-expertise model of professional organization and efficiency. Although nurses had been trying to win professional autonomy since the last third of the nineteenth century, their efforts continued to be frustrated not only by hospitals and physicians but by their own internal divisions. The needs of the rank and file for better income collided with the leadership's allegiance to the Nightingale ideal of sacrifice. Nonetheless, the medical profession's success in raising its own status under the banner of science suggested that an analogous strategy might accomplish the same for nursing. Aspiring nurses in the early twentieth century developed their own version of Frederick Taylor's extraordinarily popular "scientific management" philosophy that subordinated individual freedom to bureaucratic regulation of production. The Progressive gospel of efficiency thus found an eager audience among the nursing leadership of the Chicago Lying-in Hospital.[64]

Koch and her nursing staff enjoyed a remarkable degree of autonomy for members of their profession. Though a prominent local pediatrician, Isaac Abt, assisted DeLee to some extent, in practice both physicians supervised at a distance.[65] Under Koch's supervision and with the aid of her care guidelines, the nursing staff made many emergent management decisions independently of any physician. Her writings suggest that some were skilled even in intubation.[66] The result was a degree of autonomy remarkable for nurses of the time, a tradition that was to continue until the incorporation of resident physicians into the premature nursery many years later.

But in the end, the demands on the nursing staff forced DeLee to contract the station. In 1905, he still clung to a grandiose vision of expanding it to encompass twelve incubators once his new hospital had opened.[67] He had little trouble in his mind justifying the need for such a facility; Koch reported to the board of directors that the incubator station had been so busy that it had turned away sixteen infants, compared to a total of twenty-three it had admitted.[68] The problem was in finding a way to pay for it all. Repeated attempts to raise funds to expand the station met with little success. After failing to raise five thousand dollars in donations to enlarge the system substantially, DeLee went so far as to donate over one thousand dollars of his own money to keep it running.[69]

It eventually became clear that DeLee's expectations were wildly discordant with financial realities. Besieged on all sides, the Chicago obstetrician finally decided to reduce the station in 1908 because of what he called the "overwhelming demands made upon it." He elaborated in his annual report: "We could not supply the nurses and other necessities to properly conduct this department."[70] Henceforth, the incubator station was dissolved as a separate unit, and the hospital employed one or two Lion incubators as part of the well-baby nursery. The number of premature infants admitted to the hospital declined to a trickle.[71] By the time the first part of the new lying-in hospital opened in 1914, DeLee was no longer aggressively promoting a central role for the incubator station. Although he still sought a small station of four incubators in the new hospital, he showed little public interest in premature infant care.[72] Indeed, in 1914, the same year that the new lying-in hospital finally opened, the Chicago pediatrician Julius H. Hess created his own incubator and thereby launched a career that would mark him, instead of DeLee, as the country's leading advocate of the premature infant.

DeLee's incubator station failed because of the demands it placed on nursing, which could not be satisfied given the economic limitations attending maternity hospitals in the early 1900s. The first Chicago Lying-in Hospital in the house on Ashland Boulevard promoted a twentieth-century technological ethos within a nineteenth-century charitable institution. Despite the strident efforts of its directing physician, nurse, and their staff, its incubator station required more financial support than was possible in a hospital that attracted few paying patients. DeLee's dilemma was not unique. Before large numbers of middle-class women began to deliver their infants in the hospital after 1910, obstetricians remained heavily dependent on philanthropy for financial support. Public assistance, moreover, was far too meager to be of value. It is not clear whether even middle-class private patients could afford the kind of prolonged and intensive care required by premature infants without the assistance of third-party insurance coverage.[73] By the time that the rise of hospitalized childbirth changed this reality, obstetricians were becoming increasingly preoccupied with the care of the mother. Neonatal medicine in the process shifted from an obstetric to a pediatric specialty.

The Retreat from the Newborn

Joseph B. DeLee's story has been told in some detail as an illustration of just how far the technological paradigm could go in early-twentieth-century newborn care. The question remains why American obstetricians by and large rejected his example, even as the rise of hospitalized birth simplified premature infant care in many respects. The answer to this problem requires the difficult task of trying to generalize to obstetric practice as a whole. Published articles in medical journals, unfortunately, shed light mainly on the relatively small number of committed advocates of premature infant care rather than on the profession at large. Moreover, the retreat of obstetricians from the realm of the newborn took place with little fanfare or controversy, characterized by a decline of obstetric publications rather than a published dispute in the literature. Only after pediatricians began a concerted series of attacks on obstetric "neglect" of the newborn after 1913 did many articles explicitly address the subject.[74] By then, obstetricians had essentially completed their retreat from the care of the premature infant. Despite the limitations of the medical literature, however, there remains enough testimony to speculate with fair certainty how obstetric neonatology became derailed.

To begin with, it is clear that many maternity hospitals acquired incubators as manufacturers made them readily available beginning in the late 1890s. Although the necessity of having to construct one's own incubator had previously presented a deterrent to many physicians, by 1897 a speaker at the American Association of Obstetrics and Gynecology asserted that the device could "be easily obtained in any of the physicians' supply houses."[75] In the course of the next several years, obstetricians quietly introduced incubators into prominent maternity hospitals in cities such as New York, Boston, and St. Louis. Though few published their own articles, obstetricians nonetheless furnished the majority of the audiences participating in discussions of the device at local medical meetings.[76] Philanthropy often supported the purchase of new incubators, affirming the continuing viability of its longstanding connections with American hospitals. It acted in concert with a new factor, the entry of private patients into the maternity hospital. When Brooklyn's Low Maternity Hospital received a Lion incubator as a gift, one of its physicians, presumably eager to attract paying patients, pointed out in a local journal that the device could be used in private rooms as well as the public wards.[77] Such appeals remained unusual in 1901 but anticipated a consumerist theme promoting medical technology that would become prominent in the 1920s.[78]

Yet the published literature reveals no early attempts to emulate DeLee's model of setting aside a special section of a maternity hospital for premature infants. Physicians might place incubators in private rooms, the ward, or in a corner of the well-baby nursery. Although one of the main theoretical advantages of the Lion incubator was the possibility of connecting it to an external forced-air ventilation system, few if any physicians besides DeLee did so. New York's Sloane Maternity Hospital, for example, introduced incubators in 1897 without making any provision for a special station to house them. It obtained at least one wooden Tarnier *couveuse* as well as a Lion incubator, both of which were set up in the regular nursery or even in the hallway, which offered the next-best substitute for true outside air.

With a supply of incubators far short of demand, Sloane's physicians sometimes had to place two or three babies together in the Lion. They developed a triage system based on the infants' weights, placing only those less than four and a half pounds (2,050 grams) in an incubator, while treating larger babies in a padded basket with hot water bottles unless they deteriorated. Some infants had to be released from the incubator or transferred to other institutions to make room for others. Sloane thus offered a complete range of treatment, from the padded basket through the *couveuse* and ending with the Lion incubator.[79] If anything, Sloane was better equipped for infant care than were most maternity hospitals. The ongoing support of the Vanderbilt family and a rising population of private patients made Sloane one of the premier maternity hospitals of its day.[80]

Efforts to evaluate the incubator were as inconsistent as the range of ma-

TABLE 6.1
Survival Rates of Incubated Infants at Sloane Hospital for Women (under James Voorhees, 1897–1900) and Paris Maternité (under Stéphane Tarnier, 1880–1883), by length of Gestation (%)

Gestational Age (months)	Sloane		
	All Patients	Excluding Early Deaths[a]	Maternité
6	—	—	16
6½	22	66	36
7	41	71	49
7½	75	89	77
8	70	91	88

Source: Data from James Voorhees, "The Care of Premature Babies," *Archives of Pediatrics* 17 (1900): 341.
[a] Excluding deaths in first few hours.

ternity hospital environments that housed it. Though most physicians probably relied on their own anecdotal experience in this regard, a few did publish formal statistical case series. Sloane, once again, furnished the largest of these before the First World War. Its author was James D. Voorhees, an obstetric resident who studied the incubator in the course of developing a new procedure to induce premature labor artificially. Voorhees had a strong predilection for the Paris obstetric style, as demonstrated by both his interest in pediatrics (refined through an internship in the New York Foundling Hospital) and his extensive use of statistics.[81] He published, in fact, the only significant American statistical study of the incubator designed to be compared directly with that of Tarnier. Voorhees followed Tarnier's example of organizing the survival rates of his 106 incubator patients according to their gestational age to facilitate comparison. Following its publication in 1900, his study was cited more frequently than any other in the American literature for a quarter century.[82]

The irony was that Voorhees, having organized his results so carefully, was in the end uncertain whether or not to consider them a success. His actual results were inferior to those of Tarnier, as he openly admitted.[83] Indeed, his overall survival rate was close to 50 percent (see table 6.1) and therefore rather equivocal; its interpretation was comparable to the proverbial question of deciding whether the glass was half full or half empty. Voorhees responded by providing two sets of results in his table, one including all deaths and the other excluding those that occurred in the first hours of life. The latter results were far more impressive in both absolute and comparative terms, illustrating neatly how the definition of the target population for the incubator to a great extent determined its effectiveness. By presenting both sets of results in his table, Voorhees made the process of definition explicit for the reader—more so, in fact, than had DeLee or Budin.[84]

One suspects that when obstetricians assessed the incubator in a more casual fashion without attending carefully to how they defined their population, their results tended to be disappointing. The incubator's efficacy was always relative to whatever else might be available; it was not a cure in an absolute sense. Physicians could show it to be a success or failure depending on who they used it on. To the extent that American physicians followed Sloane's example and concentrated their resources on one or two expensive Lion incubators rather than a set of cheap Tarnier-Auvard *couveuses,* they necessarily had to reserve them for the smallest premature infants. For example, an obstetrician attending a meeting of the New York Academy of Medicine announced that nine of the eleven babies he had treated in the device had died. But the eleven incubator babies came from a larger series of forty-one premature infants.[85] In the end, statistics by themselves failed to settle the question of whether the incubator worked.

Voorhees's decision to present two sets of results, one better and one worse than those of Tarnier, reflected a deeper ambivalence within the obstetric profession regarding the care of the premature infant. These feelings in turn mirrored the difficulty of fitting newborn care into the increasingly surgical and acute-care paradigm overtaking obstetrics. In DeLee's case, this orientation promoted an aggressive and even mechanistic approach to the early complications of prematurity. But ultimately, as DeLee's experience illustrated, the acute-care paradigm worked to undercut obstetric interest in the premature newborn. As it became clear that premature infant care required sustained efforts over many weeks rather than simply the application of a new technology, it lost much of its appeal. Moreover, the increasing tendency of obstetricians to develop new interventions directed at the mother left little time for supervising the newborn. If anything, the traditional tendency of obstetricians to limit their pediatric responsibility to the act of resuscitation quite likely became stronger as their profession became distinct from general practice and midwifery.[86] The full-time obstetrician was the mother's doctor, not the infant's.

Pediatricians meanwhile were developing their own claims the special expertise necessary for the medical care of the infant. Though primarily interested in problems of older infants, by the early 1900s pediatric specialists were turning to premature infant care, as well. Obstetricians varied in reaction to the elaborate feeding systems developed by their pediatric colleagues, sometimes imitating them but more often asserting the supremacy of breast milk. Yet they did in many cases cite pediatricians in their literature and invite them to their meetings when discussing premature infants.[87] It became increasingly hard for an obstetrician caring for a premature infant living not merely for days but weeks in the hospital to avoid the sense of entering unfamiliar territory.

As had happened in DeLee's example, the premature infant ultimately became the responsibility of the obstetric nurse. Indeed, in some cases obstetricians came to practice what amounted to a philosophy of benign neglect with regard to sick and handicapped newborns. They did not so much articulate a policy as allow one area of the hospital, the nursery, to remain in a nineteenth-century state of inattention while the rest underwent modernization. Even the better maternity hospitals furnished inadequate personnel and equipment to deal with premature and other ill newborns. Essentially, all responsibility was left in the hands of an often overextended nursing staff. Mary Breckinridge, later the founder of the Frontier Nursing Service, left the following portrait of the nursery at the New York Lying-in Hospital during her student internship in 1907:

> In the nursery to which I was assigned, both for day and night duty, there were never fewer than twenty babies and sometimes as many as thirty. The heat seemed to be turned on rather low at night and the east wind, from over the river, penetrated through the cracks of the windows. Only one thin cotton blanket was allowed to each bassinet. It was cold enough at night for us to take sweaters to put over our uniforms, but I never had the heart to wear mine. It was an extra covering for at least one baby.[88]

Breckinridge did not specifically mention premature infants, who could hardly have thrived under such circumstances, but did recall an infant handicapped with spina bifida who survived for weeks under the care of the nurses, until dying when a transfer to another hospital was attempted. This description pertained to one of the leading maternity hospitals of its day, an institution that had moved, in 1902, to an eight-story building following a one-million-dollar donation from J. P. Morgan.[89] Other institutions likely fared still worse.

The other manifestation of the limited capacity of obstetricians to care for premature newborns was the transfer of chronic patients to infant hospitals. This practice, mentioned by Breckinridge, was important in shaping the pediatric perception of the premature infant. In the early 1900s, it was unusual for a pediatrician to be allowed access to the nursery; the case of Isaac Abt at DeLee's Chicago Lying-in Hospital appears to have been a rare exception. But obstetricians frequently involved pediatricians indirectly by discharging older premature infants to their care. A premature infant often overcame the special challenges of the first days of life only to succumb later to one of the great scourges of infant life, gastroenteritis and respiratory infection. Transfer to an infant hospital, a dangerous setting for an infant under any circumstances, increased the chance of mortality still further. Sloane Maternity Hospital, for example, transferred 9 of the 106 incubator babies in Voorhees's series to local infant hospitals between 1897 and 1900; by the latter date, seven of the nine had died of gastroenteritis.[90] By sending such infants to pediatricians, obstetricians helped to define two kinds of premature infants as seen by the two specialties: an acute obstetric variety subject to hypothermia and asphyxia and an older, more chronic pediatric population subject to malnutrition and infection.

It is thus relevant at this point to turn to the pediatric profession during these same years and ask how its ideas on premature infant care were evolving. While few pediatricians considered the reduction of newborn mortality to be a high priority before the second decade of the century, their earlier experience would shape their attitudes once they did gain responsibility. The fragmentation of newborn care between the two specialties meant that each would understand the premature infant in quite different ways.

❧

The incubator traced a meteoric course after contacting the American obstetric profession, one that began with a fiery explosion of new ideas that were quickly extinguished as promise gave way to reality. The most spectacular manifestation of the period was the creation, by Joseph DeLee, of the first American incubator station. Yet the relevance of this phase of development to the history of American neonatology may well be questioned, for the incubator station, like the incubator show, was in many ways an anomaly in its own time. This fact is easy to forget, given the remarkable degree to which DeLee's incubator station anticipated ideas that have become central to modern neonatology. First, DeLee appears to have been the first physician to advocate the regionalization of newborn care; his idea of developing centers for complex technological care of premature infants supported by transport systems would later become a basic principle underlying the organization of neonatal intensive care.[91] Second, he anticipated to a remarkable extent the mechanistic style of intensive newborn care so prominent today. DeLee envisioned the respiratory distress of the premature infant in the framework of asphyxia, rather than constitutional weakness, and employed both intubation and aggressive physical interventions to counteract it.

These ideas required further technological support to become practical in the twentieth century. Physicians in St. Louis, who seem to have been aware of DeLee's transport incubator, built their own at the Louisiana Purchase Exposition in 1904. Yet the limitations of having to rely on existing rail and road systems undercut the device's advantages. Less than one in six infants survived the trip to the incubator station.[92] Transport incubators dropped out of favor until the 1920s and 1930s, when Julius Hess and the Chicago Health Department reintroduced them, this time linked to automobile ambulances.[93] Mechanical respiratory support for the premature infant for the most part awaited the 1960s, in spite of some intriguing earlier experiments. For example, few physicians are aware that an early modification of the Drinker respirator (or iron lung) later used for polio victims was applied to premature infants in the early 1930s.[94] DeLee did not influence any of these later innovators but, along with many other obstetricians, did espouse a conception of prematurity in terms of asphyxia that became influential in promoting both the use of oxygen and respiratory support.

Nonetheless, DeLee's ideas did not fail simply for lack of technological support. They represented an obstetric perspective on the premature infant that declined as that profession withdrew from the newborn. Obstetricians could ill afford spending long hours with premature infants as the care of the mother became more time consuming and thus left such infants to their nurses. Transferring their care to pediatricians was far less straightforward, since the two professions often worked in different hospitals intended respectively for women and infants. Pediatricians tended to see older premature infants brought from the home or transferred from maternity hospitals. The spectrum of premature infant physiology and pathology thus was split between the two specialties. Pediatricians had few chances to gain experience involving the acute management of prematurity or to appreciate first hand the importance of resuscitation and preventing temperature loss. Indeed, with time, the activist and early-interventionist approach of DeLee came to be not merely neglected by the pediatric profession but actually resisted. The battle was framed around the role of the incubator itself, which by 1910 had taken a very different course within the pediatric setting.

CHAPTER SEVEN

The Pediatric Revolt

By 1910, obstetric interest in the premature infant was clearly waning. The specialty's most promising advocate of newborn care, Joseph DeLee, had declared a strategic retreat. The United States did not produce a Pierre Budin, an obstetrician who maintained a lifelong focus on improving newborn care. American obstetricians, as far as can be discerned from their writings, by and large delegated most of the infant's care to the nurse, much as physicians had deferred to mothers during the previous century. As a result, the invention of the incubator to a remarkable extent left intact the traditional division of responsibility between physicians and women regarding the newborn.

Pediatricians eventually moved into the territory vacated by the obstetric profession but did not simply build on their colleagues' earlier accomplishments. Instead, they brought their own set of experiences and assumptions to the care of the premature infant. Though pediatricians shared the obstetric enthusiasm for science, they framed this orientation in a medical rather than surgical paradigm. The result had great implications for the incubator. Whereas obstetricians had concentrated on developing mechanisms to maintain a stable internal environment, pediatricians were more interested in asking what that environment ought to be. Instead of assuming the value of breast milk and fresh air, pediatricians asked whether science could identify and thereby manipulate their beneficial components. The point was not so much to imitate nature but to study and improve on it.

But the aspirations of science were to collide with the far-from-ideal reality of the infant hospital setting, unleashing a backlash of doubt and dissent over the future of scientific pediatrics. Far more than had been the case with obstetrics, the incubator itself became a lightening rod in the pediatric setting for a range of more fundamental questions regarding the limits of medical responsibility for the premature infant. It triggered anxiety over the balance of power between physicians and mothers in caring for the newborn, between the hospital and home as the setting for its care, and between scientific medicine and domestic hygiene as the foundation for its management. In this setting, the symbolic qualities of the new technology assumed central importance in determining its reception.

Prelude: The Ideal of the Artificial Environment

Pediatric interest in the newborn was for the most part sporadic at the turn of the century, at least as manifested in the published literature. The specialty was just beginning to develop a formal organizational structure and professional program. In 1887, a group of forty-three physicians, mostly with a part-time interest in the diseases of children, formed the American Pediatric Society. It was an elite society limited in membership and open only by invitation; most of its members were from New York, Boston, and elsewhere on the East Coast. There was little discussion of the newborn period at its annual meetings. The society existed to promote research, which initially consisted for the most part of clinicopathological studies of specific diseases. At the same time, the newer medical sciences of bacteriology and chemistry were commanding more attention.[1] The prestige of the former reflected the worldwide excitement following the spectacular results of diphtheria antitoxin in children. But it was medical chemistry that would provide the first path leading to the newborn, through the development of artificial formulas for young infants.

Illustrating the linkage between feeding and the incubator was the first pediatrician to take serious interest in the premature infant, Thomas Rotch of Boston, whom we have already encountered in the earlier discussion of his 1893 brooder. For all its originality, Rotch's incubator was something of an anomaly in its own time. Indeed, its inventor's interest in premature infants was equally extraordinary when set against that of contemporaries. While most pediatric textbooks devoted only several pages at most to the problems of premature infants, Rotch provided a major chapter directed exclusively to their care in his own 1895 textbook. It is this chapter, initially given as a lecture to students, that provides most of our information on his incubator.[2] As a result, it is difficult to trace the subsequent history of the device, much less to construct a detailed case study. In spite of these limitations, Rotch's testimony offers some critical insights into how even a pediatrician especially sympathetic to the incubator nonetheless understood it in a way fundamentally different from that of its original inventors.

Rotch seems to have developed his unusual interest in the premature infant as an offshoot of his search for clinical problems amenable to his new scientific feeding methods. He had joined Harvard's medical faculty in 1878, one of the early members of the new generation of physicians who had gone to Germany for their postgraduate training. In physical appearance, he embodied the Boston Brahmin turned scientific expert, a short man with a full mustache and high-pitched voice, remembered for arriving at the hospital in a fancy horse-drawn carriage well into the early twentieth century.[3] His reputation rose rapidly during his own lifetime and then fell nearly as quickly after its end in 1914, in parallel with the meteoric success of the "percentage" feeding method that became indelibly linked with his name.[4]

More than perhaps any other single individual, Rotch made the supervision of artificial infant formula one of the central roles of the early-twentieth-century pediatrician. Wet-nursing had never been popular in the United States, discouraged by economics as well as by upper-class concerns that the practice could inflict physical as well as moral harm.[5] But the private infant-formula industry took advantage of this situation by promoting products of dubious safety and scientific value, from the perspective of pediatricians. In contrast, Rotch set out to modify cow's milk into a product that imitated breast milk as closely as possible. His infant formula could be obtained by prescription in large cities from special milk laboratories or prepared at home using elaborate equations mixing exact proportions of milk, cream, and sugar. Though its complexity eventually led to a backlash, percentage feeding attracted an enthusiastic following among academic physicians at the turn of the century; one went so far as to call it the "Eden of pediatrics."[6] With the aid of Rotch's equations or a local milk laboratory, artificial formula could be prescribed to meet the particular needs of an individual patient, modified if necessary on a daily basis. It was this principle of individualizing milk according to the stage of the infant's development that led Rotch to the premature infant, for Rotch saw premature and delicate infants as providing the ultimate test case where his formula would prove to be actually superior to breast milk.[7]

Rotch's brooder reflected an understanding of environmental medicine similar to that of his artificial formula. In each case, he took a product from agriculture (the incubator and cow's milk) and modified it along the lines suggested by the study of human physiology. Rotch mixed two metaphors, as we have seen earlier, in designing this environment, reflecting the ambiguous status of the premature infant between fetus and delicate infant. As an artificial uterus, the incubator shielded the baby from harmful external stimuli, including those of its own mother. As an artificial home, it controlled through its ventilation mechanisms the atmosphere respired by the baby. In both cases, science enabled the physician to improve upon Nature rather than blindly obey her.[8] Again, Rotch employed a similar kind of thinking with regard to feeding. While conceding, for example, the superior results obtained with breast-feeding, he asserted that "the latest scientific work on this subject shows very clearly that it is not breast-milk, as a whole, which is pre-eminently good, but that there are definite known reasons inherent in the different elements of the breast-milk which make it the best known food."[9] The brooder would serve to rationalize the environment much in the same way his percentage feeding method demystified breast milk.

Rotch thus used the metaphor of the artificial environment as a starting point from which to develop more specific agents to manipulate the infant's physiology. In place of the uncontrolled stimulation of the mother and home environment, he substituted specific stimulants more amenable to medical control. One of the particularly intriguing features of his incubator was its provision for connecting a tank of oxygen. This innovation at first sight suggests a remarkable anticipation of the future, for during the 1930s and 1940s physicians would transform the incubator into an oxygen chamber for treating respiratory distress.[10] But in the 1890s, physicians for the most part applied oxygen as a fast-acting stimulant. Available since the 1870s in pressurized canisters supplied in major American cities, oxygen was primarily administered in short bursts for chronic diseases such as anemia and tuberculosis.[11] Rotch did not attempt to maintain a specific concentration of oxygen in his incubators but applied the gas two or three times a day, often coupled with a second stimulant, such as brandy. The gas was given only in restricted doses, in a manner similar to the dispensation of any pharmacological agent.[12]

The Rotch brooder, introduced simultaneously with percentage feeding, thus made the infant's environment a matter of concern for scientific physicians as well as mothers. His refined concept of ventilation led him to reject the popular hygienic notion, dating back to the hardening theories of Locke and Rousseau, that fresh air had to be cold in order to invigorate the constitution. "Infants in our Northern climate," he claimed, "are exposed to cold far more than they ought to be."[13] In this respect, he set his therapeutic approach firmly apart from that of his competition, the traditional (and nonprofessional) infant nurse. One case report in his textbook is especially revealing. Rotch remarked testily how the nurse of one of his premature patients, "possessed with the idea that it needed plenty of cold fresh air," left the baby exposed to an open window during cold winter weather. The infant subsequently died.[14] Rotch admonished his students through such anecdotes to warn mothers to observe their nurse carefully, for "the idea that the child should be taken care of by an old, experienced nurse is a vicious one." In extending the domain of the physician into that of the mother, Rotch thus sought to form an alliance transcending the line of gender by appealing to the traditional divisions of class. Rather than taking the strategy of competing with the mother, he appealed to her middle-class sympathy for scientific motherhood as contrasted with the traditions of the lower-born nurse, whose experience he dismissed as consisting of "ignorance rather than of intelligence."[15]

Rotch himself, of course, would not have seen his mission in terms of class or gender. His devotion to science was both intense and genuine. There was a sense in which he almost deliberately ignored any possible limits to the potential of the artificial environment in countering the defects of heredity. While stressing a rigorous method within the laboratory, Rotch also tended to display an uncritical faith in the easy translation of such knowledge into actual practice. In contrast to Tarnier and Budin, he provided no statistical assessment affirming the superiority of his elaborate incubator to its humbler precedents. His testimony consisted entirely of isolated case reports. The implication was that the utility of a new technology could be assumed rather than proved, so long as it was based on sound science. Rotch displayed a similar confidence regarding the future of his patients. "The question is often asked," he wrote, "whether premature infants, even if their lives are saved, can be as well developed physically and mentally as are those born at term."[16] He answered the question affirmatively yet provided no evidence beyond a set of dramatic "before-and-after" pictures of an incubator baby.[17] Medical ethics for Rotch above all entailed the duty not to compromise the "science" behind scientific medicine.[18]

Rotch tended to deny economic limits as well. His technological perfectionism greatly restricted his appeal beyond the upper-class clientele of such affluent communities as Beacon Hill and Fifth Avenue. The Boston pediatrician insisted that the successful care of premature infants required employing the best possible system, "no matter what the cost."[19] Here again, the incubator exemplified the philosophy behind percentage feeding, concerning which Rotch reminded physicians that "the nutriment we are endeavoring to copy, far from being a cheap product, is, on the contrary, a very expensive one."[20] Rotch's idealism led him to propose an ambitious network of milk laboratories throughout the nation providing physician-prescribed artificial formula and renting incubators for home use.[21]

Unfortunately, the available source material sheds little light regarding the extent to which Rotch tried to incorporate the brooder into this envisioned system. He did succeed in developing milk laboratories in a number of cities for preparing formula, but he made no further mention of incubators when describing their function in later years.[22] Why he lost interest is unclear. The success of the milk laboratories themselves reached a climax at the turn of the century, leading to a more protracted period of retreat hardly favorable to expanding the system to include home incubators.[23] Without a rental system, the economic objections to the brooder were formidable for most families. One reviewer of his 1895 textbook, who applauded his incubators for being "as near perfection as can be attained," added that "their expense and elaborate construction render them, however, unavailable for most cases outside the hospital."[24] The principal objection to the use of the Rotch brooder in the home, in fact, was its expense.[25]

It is essential to grasp the element of scientific idealism behind Rotch's reinvention of the incubator in order to understand the appeal of the device, as well as its eventual downfall. The idea that science could enable physicians to create a perfect living environment for a prematurely born infant held great excitement for the young pediatric profession. Pediatrics almost by definition attracted physicians enthused over the possibilities of pushing environmental therapy to its limits, of testing how much of the "inevitable" mortality of early infancy was truly unavoidable. Their dilemma was that they worked in a milieu that offered no such freedom from outside constraint. Unless they were willing to restrict themselves to affluent city dwellers, pediatricians had to adapt the incubator to the hospital. Yet the hospital of the late nineteenth century remained a dangerous and threatening place for a young infant. The introduction of the incubator into the infant hospital environment brought about another sharp collision of scientific ideals and social realities.

Premature Babies and the American Infant Hospital

A short anecdote may suffice to introduce the problems of caring for premature babies in an infant hospital. One of the first recorded instances of an incubator used in an American hospital appeared in a short human-interest story in the *New York Times* on 1 February 1891. It detailed the struggle of Baby Johnson, whose story began amidst circumstances all too familiar to the residents of a city in which as many as one baby in five may never have lived to see a first birthday.[26] Born two months early to a mother who survived her child's birth by less than four days, this unfortunate "mite of humanity" survived his first month of life only through the unremitting care of a family friend. But the resources of the home eventually were exhausted, and the two-pound infant was brought to the New York Babies' Hospital cold and emaciated, one lung in a state of partial collapse. It was here that this commonplace tragedy converged with the American pediatric profession, for the hospital's physician was L. Emmett Holt, a rapidly ascending young physician who was to play a central role in the transformation of American pediatrics into a profession. The baby was placed in an incubator the hospital had just installed. The nurses were initially optimistic, commenting to reporters that the infant might grow up one day to become president of the United States.[27] But within two days the baby died. Holt speculated that the infant would likely have survived had its caretaker not waited a month before bringing it to the hospital.[28]

As experience would later demonstrate, Holt had run into a barrier that would frustrate an entire generation of pediatricians employing the incubator. Why were mothers and caretakers so reluctant to hospitalize their infants? The answer had to do with how mothers understood the infant hospital. For infant hospitals were, quite literally, the progeny of foundling asylums, institutions for abandoned rather than sick children. An understanding of the connection between the two is essential to the task of unraveling the pediatric response to the incubator.

Compared with their European counterparts, foundling hospitals in the United States were comparatively young institutions created by philanthropists in the 1860s and 1870s.[29] They represented a response to the problems of high infant mortality and child vagrancy that plagued eastern seaboard cities in the wake of the explosive growth of immigration after the Irish potato famine. New York City, whose slums were more than twice as crowded as their densest counterparts in London, stood at the heart of the movement. Beginning with the creation of Nursery and Child's Hospital in 1854 by the philanthropist Mary Dubois, a series of infant asylums and foundling hospitals came into existence to provide institutional care for unwanted and abandoned children. Private subscriptions supplemented by a city public appropriation supported these efforts. Few admissions were truly anonymous "foundlings"; Nursery and Child's Hospital, for example, was sensitive to the issue of encouraging abandonment and frequently admitted mothers along with their children.[30] But New York parents did abandon their infants to these institutions in astounding numbers. The New York Foundling Hospital, especially, resembled its continental namesake, admitting abandoned infants to be distributed to a system of home wet nurses. The Catholic Sisters of Charity, who opened the institution in 1869, initially planned on taking fifty to one hundred infants during the first year. Instead, some twelve hundred abandoned infants poured into the institution, forcing it by 1873 to dispense with its anonymous admission policy and move to a larger building.[31]

One of the great objectives of the pediatric profession was to tame these institutions and thereby harness them for the benefit of medical science. The process proved to be long and arduous. Women dominated the early infant asylums, providing most of their staff, including the nurses, board members, and physicians. Male pediatricians did not fit easily into this milieu, in spite of their desperate need for facilities for research and training. The early pediatric leader Abraham Jacobi openly criticized the staff of Nursery and Child's Hospital for the institution's high mortality rate, an offense that led to his resignation in 1873.[32] But following the creation of the American Pediatric Society (APS) in 1887, pediatric specialists began to seek a more active role in these hospitals. Foremost among them was L. Emmett Holt, perhaps the leading American pediatric authority of the early twentieth century and the author of his generation's most influential textbook, *The Diseases of Infancy and Childhood.* Holt's expertise derived largely from his clinical and pathological experience obtained while medical director of the New York

Babies' Hospital from 1889 until his death in 1924.[33] Elected president of the APS in 1898, one year after the publication of his textbook, he used his presidential address to expound on the necessity of infant hospitals for research, teaching, the training of nurses, and patient care. Infants would never be distinguished from adults, he argued, until they could be studied and treated in their own special setting.[34]

Holt's address to the APS marked the beginning, rather than the culmination, of infant hospital reform. Well into the early twentieth century these institutions remained in transition, retaining many features of the older foundling home while anticipating the modern children's hospital. At the New York Babies' Hospital, the staff remained overwhelmingly composed of women, ranging from the pathologist Martha Wollstein to the resident physicians and nurses. Young girls composing the "Cribside Committee" daily ministered to the patients, according to one account, "in many loving ways, which even babies appreciate."[35] The medical board did not vote to allow male interns until 1911.[36] Women imparted a maternal, protective, and religious tone to the institution. Indeed, sun gardens and playgrounds, high ceilings and fireplaces, comfortable furnishings, and summer seaside convalescent homes were standard fixtures of infant hospitals at the turn of the century. There was little evidence of technology beyond the sterilizing apparatus used to prepare milk. These aspects of institutional life pointed back to nineteenth-century notions of the proper moral and sanitary environment for the rearing of vulnerable children.[37]

Contrasting with this serene and even pastoral setting was a class of patients whose mortality mirrored the surrounding world of the tenements. The phenomenon of high institutional mortality, sometimes termed "hospitalism," continued to afflict infant hospitals, just as it had the foundling home.[38] Although both the opponents and the defenders of infant institutions manipulated statistics to suit their purposes, even in the best-regulated institutions roughly half of all infants under a year of age died following hospitalization.[39] The largest single cause of this mortality was a pattern of malnutrition and recurrent infection Holt called "hospital marasmus." It afflicted not only babies who were already severely malnourished on admission but also many apparently robust infants brought to the hospital for an acute illness who then mysteriously began to lose weight. "I have seen scores of infants who were plump and healthy on admission lose little by little," Holt asserted, "until at the end of three to four months they had become wasted to skeletons—hopeless cases of marasmus, dying of some mild acute illness such as an attack of indigestion or bronchitis."[40] Holt could not explain the condition in terms of nutrition, nursing care, or defective hygiene. He performed over 200 autopsies in his quest to solve the problem but was unable to demonstrate any consistent findings of importance.[41] It seemed that somehow the hospital environment itself was at fault. In 1898, Holt told the APS that the phenomenon of hospital marasmus represented the greatest obstacle to the future of the infant hospital; indeed, it demonstrated "beyond any question how injurious to infant life is the atmosphere of a hospital."[42]

This context of high institutional mortality profoundly shaped the response of the pediatric profession to the incubator. The benefits of a new therapy could be masked by the mortality induced by the hospital; indeed, a given disease process could be shaped by that environment. Some of the poor results from early diphtheria antitoxin trials, for example, may have reflected the additional morbidity contributed by the hospital setting itself.[43] A premature baby in an infant hospital presented a different set of problems and concerns from one born at home or in a maternity hospital. Its very distinctiveness could be lost against the background of hospitalism. Before the introduction of the incubator, premature infants admitted to foundling institutions most likely were lumped together with the broader undifferentiated category of "weaklings." A physician at the New York Foundling Hospital, where most admissions were only a few days old, described the marasmic infants dying at his institution in terms that could easily encompass premature birth: "Some are moribund when left to the care of the hospital. Many have suffered from ignorance, neglect, or exposure. Not a few are the subjects of constitutional disease. These weaklings, whether sent out to nurse or kept within the institution, soon take their places in the marasmus ward and go to swell the death lists."[44] The entity of hospitalism added a new dimension to the age-old notion of the "delicate" infant with which the premature infant had always been identified. As a consequence, it shaped the pediatric response to the incubator.

The Incubator under Fire

While source materials are too limited to allow us to assess the response of pediatricians in general to the incubator, we can at least gauge the range of their responses by focusing on a select number of examples. The premature infant simply did not constitute enough of a medical or public health imperative during the early 1900s to lead the American Pediatric Society or the American Medical Association's pediatric section to try either to set standards of practice or to measure them in the community. The story we are undertaking to describe is of interest not because of its significance to contemporaries but because attitudes during the early 1900s survived to shape practice in later years, when the premature infant finally did become a priority. In order to measure opinion before 1910, we will therefore center our attention on two institutions that exemplified the ideological poles of professional pediatrics. In particular, they were led by individuals who took different positions regarding the relationship of science to pediatric practice, Thomas Rotch and Henry Dwight Chapin.

The examples of Rotch and Chapin illustrate how the banner of "scientific pediatrics" could provide the young profession a unifying language that masked disagreement over just what "science" actually meant. Chapin, for example, appears to have been one of the first pediatricians to apply the Rotch incubator to the hospital setting. Yet he was drawn to the device, and to the premature infant, for very different reasons. In contrast to Rotch's focus on pediatrics as applied laboratory science, Chapin viewed pediatrics equally as an application of social science. He saw the child, in other words, as providing an unrivaled opportunity for studying and pushing environmental reform to its limits. To that end, he became director of a hospital pediatric service (the Babies' Wards of the New York Postgraduate Hospital) to obtain a window on the poverty of Manhattan's lower East Side.[45] Chapin's research methodology accordingly differed from Rotch's as well; he gravitated toward statistical studies of populations rather than laboratory-based studies of individual cases.[46] Perhaps not surprisingly, Chapin became an early opponent of Rotch's percentage feeding method, arguing that infant milk could not be reduced to a matter of chemistry.[47] Of more immediate interest is that he soon reversed his support of the Rotch brooder.

In a presentation to the pediatric section of the New York Academy of Medicine on 9 November 1899, Chapin became the first prominent American physician to attack the incubator directly. He recounted the statistics of his first series of seventy-three incubator babies, including forty treated with the Rotch brooder. Only two had survived. The meeting's reporter noted somewhat dryly that Chapin "could not speak with much enthusiasm" for the apparatus." In keeping with his sensitivity to the effects of the social environment on the individual, Chapin added that most of the infants had arrived in a "moribund" state following prolonged exposure to the wretched and crowded atmosphere of the tenements. But he also condemned Rotch's complicated incubator itself for not deriving its air supply from outside the room. The implication was that the confined space of the incubator presented the same danger to its occupant as did the tenement.[48]

The comments of Chapin and others at the meeting illustrated how the discourse over hospitalism shaped the ways in which pediatricians understood the problems of prematurity. The premature infant population of the infant hospital recalled that of the French *service des débiles* in being characterized by deterioration prior to admission. Where American pediatricians differed was in their perception that premature infants constituted the extreme of a far larger population of infants for whom the hospital also constituted either a last resort or a place to die. An account written by Chapin soon after he first acquired his incubators revealed how similarly he envisioned the problems of prematurity and inanition, a medical term referring to the dramatic malnutrition that could characterize hospitalism in young infants:

> Many seven and eight month babies have been kept alive for weeks, but they usually die at last from inanition. The cause of this appears to be two-fold; first, they are not received in time, many being cold and blue when brought to the hospital after one or several days' exposure in a tenement house; second, their organs do not appear to be sufficiently developed to maintain continuous life. The incubator always prolongs life, and doubtless under favoring conditions will not infrequently save it.[49]

Two points emerge from these comments. Chapin saw a population that he believed was damaged not simply by cold but by the broader insult of exposure to the tenements. In addition, his infants often lived for a considerable time only to die from malnutrition. This late mortality seems to have disturbed Chapin as being the most preventable, leading him away from the peculiar problems of premature infant physiology to their overlap with hospitalism. Another physician at the 1899 meeting, Vanderpoel Adriance of Nursery and Child's Hospital, showed a similar kind of thinking in grouping premature infant deaths into two roughly equal groups, one consisting of early problems characteristic of prematurity (such as cyanosis) and the other of the later mortality from malnutrition, diarrhea, and infection. "These deaths," he wrote of the second group, "can be ascribed to the usual conditions which threaten the life of any infant in an institution."[50]

It should not be assumed that pediatricians such as Chapin were oblivious to the ways in which their social environment influenced their experience with the incubator. They were well aware of the superior results obtained by maternity hospitals in France and the United States. The obstetrician James Voorhees, for example, presented his more encouraging statistical series from the Sloane Maternity Hospital at the 1899 meeting; Chapin continued to cite his figures ten years later. But while conceding that this study demonstrated the advantage of the incubator in maternity hospitals, Chapin questioned its relevance for the context of the infant hospital, where patients had access to neither immediate care nor breast milk.[51] Indeed, Chapin recognized that the great variation characterizing the patient populations of different settings—maternity hospitals, infant hospitals, and private practice—subverted the use of simple statistical series as an objective means of comparison.[52] Though he called for more rigorously defined patient populations in the future, in practice anecdotal experience became the chief factor determining the evolution of neonatal care for many years to come.

Though Chapin henceforth published no more papers on the incubator, he continued to address the problems of premature infant care within the hospital. His response in some ways recalled the maternal style of the French obstetricians. After his long string of deaths using the incubator, he finally succeeded in raising two infants using breast milk and a padded basket.[53] Breast milk, he concluded, had to be more important than the incubator. Chapin also shared with Budin a conviction in the importance of involving the mother as opposed to using a wet nurse; he had long advocated relatively liberal visiting hours and employed a social worker to visit the homes of patients following discharge.[54] In this regard he took the remarkable step of creating not a premature infant nursery but a special ward equipped with incubators and cots that could admit mothers along with their infants. The hospital's annual report in 1907 provided an account of a tiny premature infant named Little Katherine brought to the hospital in a soap box by a neighbor. The infant was first placed in an incubator;

> As soon as possible, the mother was brought in and given a bed in that beautiful memorial ward for 'mothers and babies,' which has saved many lives, in order that natural feeding could be given. . . . After some weeks, already so smiling that she had become *Kitty* instead of *Katherine*, baby left the wards weighing four pounds, in a very healthy condition and able to make quite as much noise as the ten pound babies near her.[55]

The account affirmed a central role for the mother, implying the infant required her presence as much as the incubator.

Chapin probably hospitalized more premature infants during the early 1900s than did any other pediatric leader, yet he gradually lost interest in the face of the failure of his innovations to reduce their mortality significantly. Given his statement at the 1900 meeting, he showed remarkable persistence at the task. Between the time periods 1896–1900 and 1906–10, the number of premature infants he admitted annually nearly tripled (from 66 to 196) while overall admissions increased by only a third (see table 7.1). Overall mortality did fall, but only relatively, remaining around 78 percent during the entire first decade of the century. Though Chapin discarded the Rotch brooder, he continued to experiment with new incubators; as late as 1909 he acquired two new devices (perhaps Lion models) that he initially hoped would provide adequate heating and ventilation. But the problem of breast-feeding in an infant hospital continued to subvert his efforts. The mother-infant ward in particular had only limited success. In his 1910 annual report, Chapin, noting that "we have probably more premature infants sent to us for treatment than any other institution in the country," credited the few who had been saved to the aid of a wet nurse.[56] Few mothers, it appears, were willing

TABLE 7.1
Premature Infant Admissions and Mortality Rates, New York Postgraduate Hospital and Boston Infant's Hospital

	1896–1900	1901–1905	1906–1910
New York Postgraduate Hospital			
Number of patients	66	157	196
Number of deaths	64	120	153
Mortality rate (%)	97	77	78
Boston Infant's Hospital[a]			
Number of patients	34	54	—
Number of deaths	21	30	—
Mortality rate (%)	62	55	—

Source: Data from New York Postgraduate Hospital, *Annual Reports of the Directors of the New York Post-Graduate Hospital, including the Babies' Wards* 12 (1896)–26 (1910); West-End Nursery and Infant's Hospital, Boston, *Annual Reports* 14 (1896)–23 (1905).

[a] Data for 1906–10 not available.

or able to enter the hospital to nurse their infants. The reasons for this are not clear, since we do not have access to their testimony. It is worth underlining, however, that almost all of these infants came from extremely impoverished circumstances, and many of their mothers were either employed or responsible for other children.[57]

An alternative to Chapin's emphasis on breast milk was exemplified by the Boston Infants' and Children's Hospitals, the institutions associated with Thomas Rotch. Though sharing the same emphasis on nutrition over the incubator, Rotch's colleagues focused on his milk laboratory feeding methods. In fact, this interest set the foundations for the first American research program directed at the premature infant. Rotch's own interest in infant feeding no doubt set the tone for this program, but the specifics were left to his associates. Charles Townsend, later to be cited for seminal work regarding vitamin K deficiency, presented a paper to the APS in 1901 on the artificial feeding of a premature infant.[58] Considerably more sophisticated was a 1904 paper by John Lovett Morse, who would later succeed Rotch as professor of pediatrics at Harvard, on the caloric needs of premature infants. Replete with citations from the German literature and published in the prestigious *American Journal of the Medical Sciences,* Morse's study was virtually the only American laboratory-based examination of premature infant nutrition before 1914. He took advantage of the known concentrations of carbohydrate, fat, and protein in his formula to calculate its caloric content, which he then correlated with infant growth curves. Premature infants, he concluded, required more calories for their size than did term infants.[59] Morse's strategy of using growth curves to establish energy demands anticipated a central principle of neonatology today; in his own time the only physician who came close to matching his work was John Zahorsky at the St. Louis fair.[60]

Nonetheless, it was not long before the Boston pediatricians followed Chapin's example in retreating from the incubator. Morse, noted for his devotion to Rotch's percentage feeding method, had given up on the device by 1905. The Boston Infants' and Children's Hospitals, in fact, had already begun returning to padded baskets soon after 1900. In the process, their combined premature infant mortality rates for the five-year periods before and after dropped from 80 percent to 58 percent.[61] Morse, like Chapin, concluded that the incubator provided defective ventilation. He asserted in a 1905 paper to obstetricians that he had never seen an incubator that could provide both a constant internal temperature and a sufficient supply of "pure, fresh, warm air"; lack of the latter resulted in a "diminution in the baby's vitality and in its resistance to infection." Morse favored the use of a padded basket with hot water bottles, which he believed offered the further advantage of providing air to the lungs at a lower temperature than that of the body.[62]

Thus, in Boston as in New York, the incubator fell out of favor as a result of circumstantial evidence. To many observers, the fact that premature infant mortality fell after the device was discontinued made a powerful case against it. The American experience seemed to contradict that of the French under Tarnier. Only a few contemporaries, however, suggested the need for more rigorous studies to reconcile these conflicting results. One physician who did so was yet another of Rotch's associates, the Harvard pediatrician Maynard Ladd. In 1910, Ladd presented to the APS a thoughtful analysis of the experience of the Boston Infants' and Children's Hospitals over the previous fifteen years. While acknowledging that the hospitals' premature infant mortality rate had declined after the incubator's rejection, he nonetheless pointed out that it still remained higher than Budin's at the Clinique Tarnier. Most significantly, the difference remained substantial, even when the populations were adjusted for weight on admission; the mortality rate of infants in the 1,500–2,000-gram range, for example, was 44 percent, as opposed to 13 percent in the Clinique Tarnier.[63] Ladd believed that the foremost reason for the discrepancy was the Boston emphasis on artificial feeding. He was remarkably circumspect on the incubator itself, noting that the hospital had neither tried the newer, better-ventilated models, such as the Lion, nor made any provision to protect the infant from the immediate danger of exposure after home birth. "The best of hospital nursing and feeding," he proclaimed, "is made useless in the end simply by the journey to the hospital."[64] Ladd's recognition of the importance of transport was unusual in the pediatric profession. Indeed, his own colleagues in Boston do not appear to have shared it, for his plea for a special premature infant unit and transport system fell on deaf ears when the new Children's Hospital opened in 1914.[65]

As much as American pediatricians might differ with regard to breast milk versus artificial formula, virtually all tended to view the special needs of the premature infant within the framework of the older, malnourished infant. Again, pediatricians rarely saw premature infants in the first hours of life; most of the infants admitted to Boston Infants' and Children's Hospitals were born at home, with the remainder transferred after two weeks of life from the Boston Lying-in Hospital.[66] An unusual mortality curve characterized this situation: death rates were lowest for infants weighing between fifteen hundred and eighteen hundred grams and higher on either side of this range. The smaller babies' mortality doubtless reflected immaturity. The larger ones experienced higher mortality, according to Ladd, because parents admitted only those who were very sick; most arrived after two or three weeks of home therapy had clearly failed.[67] They suffered from problems more related to hospitalism than those specific for prematurity.

By concentrating on the second half of the older designation "premature and feeble infant," pediatricians thus emphasized the aspects of therapy that overlapped those for with all weak hospitalized infants. Holt's triad of "care, fare, and air" for hospitalism guided pediatric management of premature babies.[68] Chapin and Morse both agreed on the primacy of these principles, as opposed to specific therapies such as the incubator. And to the extent that published literature reflected pediatric opinion, their convictions were widespread. Pediatricians tended to write articles on the "care and feeding" of premature and weak infants rather than on the incubator itself. Though an occasional individual such as Maynard Ladd might call attention to the problem of transport, virtually no pediatric hospitals developed such a system to address the issue during the early 1900s.[69]

Underlying this overall consensus was a tension between scientific and domestic hygiene in fashioning the principles of premature infant care. Chapin and Rotch might both recognize nutrition to be more important than the incubator but still parted ways over the choice of artificial formula or breast milk. Similarly, there were questions over who should provide nursing care, and how. For Morse, as for Rotch, premature infant care was not a matter of common sense or good intent. Recalling his mentor's intrauterine analogy, he asserted that such delicate babies should be handled as little as possible; no one but immediate family should see them, and even they "should be allowed but one look."[70] Chapin shared respect for professional nurses, but the existence of his mother-infant ward suggested a greater willingness to include the mother as well as the nurse. Still, there was little discussion of the mother's role during the first decade of the century. The issue that became most contentious after nutrition was the third arm of the hygienic triad: air.

Propelling the ventilation issue to the forefront was a remarkable revival of fresh air therapy that swept the United States during the early 1900s. As already discussed, there was always a question over the extent to which the benefits of fresh air could be harnessed by science and technology. Ventilation offered a subject for scientific investigation but simultaneously raised all kinds of connotations drawn from the deep currents of American pastoralism.[71] If anything, the latter acquired new momentum during the early twentieth century. The critical development in this respect was the rise of a national tuberculosis campaign shortly after 1900 that promoted and popularized fresh air as never before.[72] Its success coincided with a disillusionment among child welfare advocates over whether nutrition alone would provide a panacea for infant mortality.[73] As a result, fresh air became a new weapon for the pediatric armamentarium, an antidote for a variety of ills ranging from pneumonia to general weakness and debility. Every summer, ailing Chicago infants were taken to fresh air tents, New York babies were sent to fresh air wharves, and Boston's youngest patients literally floated out to sea aboard the city's fresh air "floating hospital."[74] By 1908, even physicians at the American Pediatric Society were speaking of fresh air with an enthusiasm reminiscent of a tent revival more than a scholarly meeting. The society devoted a special session to discussion of fresh air's role in pneumonia, malnutrition, and a variety of other ills.[75] "Let us all join hands and preach fresh air," one speaker enjoined the crowd. "Vote for open squares, endorse roof gardens, have adenoids and tonsils removed . . . and work for the fresh air treatment with the same zeal and enthusiasm we have worked for fresh and pure milk."[76]

What made fresh air relevant to the premature infant was that it was also emerging as an antidote for hospitalism. A new wave of enthusiasm for roof gardens and country hospital branches spread through the profession in parallel with the fresh air movement. Philadelphia's Edwin Graham related to the APS a dramatic anecdote of how, convinced that "the so-called hospitalism was nothing but a lack of fresh air," he had set his convalescent patients outside for two hours daily on the fire escapes. The patients' strength and overall condition improved dramatically.[77] Yet it remained for Henry Chapin to pursue the implications of fresh air therapy to its logical end and conclude that the only cure for hospitalism was to move the patient out of the hospital. His presentation to the 1908 APS meeting discussed the progress of his Speedwell Society, an organization he had set up six years earlier to send convalescent and institutionalized infants to the country, where they would benefit from "plenty of fresh air, good general hygiene, and individual care." He attributed his excellent results as much to the pastoral environment as to the individual care provided by the foster mother.[78] This paper marked an important turning point in his own career, which would increasingly carry him away from the hospital and the pediatric mainstream and more to the side of nonmedical infant mortality reformers.

Given this almost millennial sense of excitement regarding the future of fresh air, it was remarkable that pediatricians struggled as long with the incubator as they did. The infant hospital experience alone suggested that the device was superfluous at best and quite possibly dangerous. The mortality rates in both Chapin's and Morse's institutions actually dropped as the device fell into disuse. The fact that these rates continued to fall after the incubator was discarded suggests the presence of other confounding factors that escaped documentation, such as the characteristics of the population admitted. Still, pediatricians were more than likely correct in suspecting that the incubator offered little for the patient in this kind of setting. The ventilation critique drove the proverbial nail into the incubator's coffin. It encouraged pediatricians to move from condemning the use of the device specifically in the infant hospital to denouncing it in general. And in many cases, it encouraged them to do so in a language that extolled the virtues of a domestic and natural environment over the hospital.

At any rate, during the second decade of the century pediatric opposition to the incubator hardened into dogma. New review articles characteristically dismissed the value of the device without offering any supporting evidence. Most advocated a return to the padded basket. One reviewer in 1912 commended this simple means of providing warmth as having "been noted by many pediatrists to be far superior to many of the elaborate incubators."[79] Another wrote in 1917: "Incubators are expensive; they are complicated. It is inconvenient to change the baby's clothing while it is in an incubator, and, most of all, an incubator is difficult to ventilate and keep free from germ contamination."[80] By 1919, another review confirmed that "the use of the incubator is becoming more and more unpopular."[81] And the Children's Bureau began by this time advising mothers that "incubators are not now generally used even in hospital cases."[82]

The continuing value of the incubator as a symbol of medical responsibility for the newborn underlay many of these criticisms, for they were often made by pediatricians who had decided that care of the premature infant was, after all, best left to the mother. The sentiment expressed by a St. Louis physician who testified that "anyone who knew the effects of hospitalism would understand his reason for being biased against the incubator" continued to link the device with the ongoing controversy over the boundary between hospital and home care.[83] Tellingly, the discussion following a 1914 AMA meeting on the subject of "The Management of Delicate and Premature Infants in the Home" featured speaker after speaker denouncing the incubator.[84] Among them was Henry Chapin, who in the same year had raised the ire of the profession through a paper entitled "Are Institutions for Infants Necessary?"[85] By this point, he seems to have lost virtually all specific interest in premature newborns, writing instead on the management of "feeble" infants.[86] He also had given up completely on the incubator. "What we gain in one factor, that is, heat, we lose in a vital factor, and that is fresh air," he asserted; "I think incubators should be abandoned entirely."[87] The pediatric profession appeared to be engaging in a full-fledged revolt against the new technology.

The Return of the Artificial Environment

Chapin issued his sweeping denunciation of the incubator in 1914, only a few months after the death of its earliest American enthusiast, Thomas Morgan Rotch.[88] Yet it was too early to write an obituary for Rotch's ideal of the artificial environment. What had invested the incubator with so much attention was its status as an anomaly within the infant hospital. As virtually the only machine on the open ward, it stood out against the hygienic and natural environment characteristic of turn-of-the-century infant hospitals. Initially a symbol of the promise of science, it eventually came to represent to its opponents an environment more akin to the tenement. Its linkage to hospitalism, in fact, raised a challenge to the idea of the infant hospital itself. Both the hospital and the incubator could be seen as exemplifying a kind of scientific hubris; their high associated deathrates demonstrated what happened, to put the issue crudely, when science tried to fool Mother Nature. Yet the infant hospital retained powerful defenders who developed a new reform program to save the institution as a base for pediatric research. In the process, the incubator was not so much discarded as absorbed into the architecture of the hospital itself.

The quintessential pediatric solution to the problem of premature infant hospital care in the early twentieth century was the incubator room. This was a literally room-sized incubator (*Monsterbrutapparate,* in German), measuring eight to twelve feet on a given side and large enough to house several infants. The idea had originated in France during the mid-1880s but attracted little attention until the prevention of airborne infection became a focus of controversy during the following decade.[89] During the early 1900s, it reappeared as an alternative to the Lion incubator, first in Italy and then in a number of major German hospitals. The incubator room was in a sense a giant

Lion, for it was a sealed environment ventilated with outside air. There were typically provisions for an adjoining heated chamber in which the nurse could feed and clothe the infant.[90] The device, in theory, solved the problem of ventilation as well as did the Lion, while its larger size created a more stable environment less prone to sudden temperature fluctuations.

In the United States, L. Emmett Holt became, in 1909, the first American pediatrician to open an incubator room. Although perhaps the most respected pediatric authority of the early twentieth century, Holt was slower than his colleagues Rotch and Chapin to take specific interest in the premature infant. Indeed, the problem may well have been forced on him. Although his institution, the New York Babies' Hospital, rarely admitted premature infants before 1903, for reasons that remain unclear the number of such admissions rose sharply during the following two years, to the point that fifty arrived in 1905.[91] Holt appears to have been both frustrated and perplexed by the arrival of so many premature babies, most of whom arrived after a period of several days in what he called "hopeless" condition. The resulting story recalled Chapin's experience. He wrote in his 1904 annual report that "in spite of the most unremitting care and constant attention, including a new and improved [Lion] incubator, only one child survived" out of thirty.[92] During the following several years, Holt continued to try to treat the growing premature baby population with a variety of means on the open ward. Sharing the concern of his colleagues over ventilation, he replaced the incubator with an electric heated pad and succeeded in saving eight out of fifty babies. Holt acknowledged that his improved results likely reflected the availability of breast milk through a hired wet nurse, but he still believed that the incubator had been part of the problem.[93] At any rate, the mortality of his premature patients remained so high that it played a significant role in preventing him from lowering his overall hospital mortality rate. In 1908, he appealed to his benefactors for a special ward for the premature babies, who by this point were being "brought to the hospital in such numbers that suitable provision for them must be made."[94]

Holt's decision to create an incubator room reflected a newer trend in pediatric hospital architecture from open wards into partitioned special divisions. Indeed, the New York Babies' Hospital under his direction became a model institution for the rising generation of pediatric specialists. Reacting to a growing concern over the importance of isolating asymptomatic "carriers" of virulent infections, Holt began culturing in 1909 all new admissions and placing them in single-crib alcoves divided by five-foot partitions, each attended by a separate nurse.[95] His reforms marked the beginning of a critical transition in infant hospital design, which would henceforth be judged increasingly by the aseptic standards of the operating room rather than the hygienic standards of the country home.[96] The new hospital would also feature more specialized divisions, for diagnosis (X-ray suites and laboratories), surgery (operating and recovery rooms), and environmental therapy (cold- and warm-air rooms).[97] It was with respect to the final division that the incubator room found its niche; Holt wrote, following its introduction in 1909, that the hospital now had three temperatures for treating feeble infants: the ordinary temperature of the ward, the cold air of the roof for patients with pneumonia and malnutrition, and the heat of the warm room for premature and very delicate infants.[98] At the same time, his incubator room was far more than simply a warm room. It was inspired by both the older emphasis on airborne infection and the newer rise of the contact theory; the room was separately ventilated with outside air but subdivided into discrete cubicles to house five different cribs.[99]

As ambitious as Holt's incubator room was, it still suffered from the inherent shortcomings of the infant hospital in the care of newborn infants. Once again, the problem of exposure prior to admission subverted the project. Holt kept the incubator room at a relatively high temperature of 90°F and initially envisioned directing it at the many premature infants who arrived at the hospital, on a nearly daily basis during the winter, with body temperatures three or four degrees below normal.[100] As the reader will no doubt suspect by this point, the incubator room failed in this setting. Holt treated successfully only ten of the fifty-eight premature infants admitted to the facility prior to 1912, with most of the deaths occurring within twenty-four hours.[101] What is more interesting is that these statistics did not threaten the concept of the incubator room the way they did the incubator.

The incubator room survived not because it produced superior statistics but because it dealt with a medical problem of increasing importance (the rising number of premature infant admissions) in a manner that accorded with modern hospital ideals. It was not an anomalous machine on an open ward but an integral part of the hospital itself, and regardless of its actual clinical results, pediatricians valued it for creating an environment that made further study possible. We may fault physicians for relying on anecdote rather than formal statistical evaluation, but it is difficult to see how the latter could have been applied to the open-ended system represented by the infant hospital. Holt and his pediatric colleagues drew their patients from a kind of black box, their status at admission shaped by multiple variables including length of gestation, age after birth, nutritional status, temperature, and infection. From the perspective of pediatricians, the problem of treatment was so formidable that developing an organized health care system, along the lines of DeLee's regional incubator stations, was premature. By carefully observing the few, Holt argued, it would be possible to determine how to treat the many. "The best results are to be obtained," he asserted in a defense of the infant hospital in 1913, "by the intensive method of study, not the statistical method."[102]

For most pediatricians during the decade after 1910, the incubator room represented the standard of hospital premature infant care. The most sophisticated examples of this approach by 1922 were the incubator rooms set up by Washington University in St. Louis, the University of California at San Francisco, and Michael Reese Hospital in Chicago. These offered such features as double-door entrances to exclude infection, anterooms for the nurses, glass partitions separating individual cribs, forced-air ventilation, and the automatic regulation of heating and humidity.[103] Relatively few hospitals could afford such elaborate facilities; some, such as New York's Bellevue Hospital, settled instead for improvised warm rooms.[104]

Yet the incubator room remained important for preserving the ideal of the artificial environment after the demise of the incubator. Though it made little provision for individualized management of a given infant's temperature, its size did offer advantages in maintaining a stable environment in terms of ventilation and humidity. The incubator room accorded well, in other words, with the values of contemporary pediatricians, who tended to subordinate temperature to the other aspects of nutritional and environmental therapy. Most of the participants in a 1921 pediatric discussion on the incubator room explicitly affirmed nutrition to be of far greater importance than temperature control; they consequently recommended only slightly elevated temperatures in the incubator room, of 75–80°F, and showed little interest in closely monitoring the infant with a thermometer.[105] In later years, humidity would become a factor emphasized equally along with temperature. Increasingly, nursing routines placed equal weight on the hygrometer and the thermometer in monitoring the room. Though humidity never really came into its own as a key ingredient of premature infant care until the 1930s, it is worth noting that a Boston study in 1933, under Morse's successor chairman Kenneth Blackfan, provided the critical impetus for its resurgence.[106]

In spite of continuities with the incubator, the rise of the incubator room symbolized the departure of the American pediatric profession from the influence of Tarnier and Budin in two important respects. First, it dispensed with the principle of individualization, the idea that premature infants needed to be kept not at a standard temperature but at one adjusted according to the body temperature and clinical signs of the given infant. Pediatricians had come to see nutrition as having far more importance than temperature and hence paid little attention to the latter. This lack of awareness had to do with the limited exposure of most pediatricians to the critical early hours of life, in which temperature played the most crucial role.

Second, the incubator room mirrored the infant hospital context by separating the premature infant from the mother as well as from other patients. It was this characteristic of incubator rooms Budin had found most objectionable. "It is better by far to put the little one in an incubator by its mother's bedside," he commented in response to the approach. "If the nurse be negligent, the mother does not fail to remark that the incubator is being allowed to grow cold." Bedside incubators also allowed breast-feeding, so that by the time of discharge "not only will the weakling have been saved, but a suckling mother will also have been conserved to it."[107] Here lay the heart of the perceptual division separating the obstetrician and pediatrician: one cared for the mother and infant, the other for the infant alone. Pediatricians were not characteristically hostile to the mother; on the contrary, Holt refused to attribute the unfortunate condition of so many of his admissions to simple maternal neglect.[108] If anything, the creation of a visiting nursing service encouraged a more sympathetic understanding among the hospital's staff of the social conditions endured by many mothers.[109]

Nonetheless, pediatricians typically found themselves responsible for treating an infant apart from its mother and consequently envisioned its care in terms of nutrition and environment. In responding to the debate over heredity versus environment that pervaded the discourse of Progressive reform, they by and large emphasized the environment. Pediatricians thus sought substitutes for the mother (the nurse and wet nurse) rather than actually involving her. What the infant needed was not so much maternal love as individualized care, nutritious food, and a healthy environment.

In comparing the French and American approaches to the incubator, one is tempted to describe the former as social and the latter as technological. The most striking features of the French approach were its emphases on the mother and the long-term outcome of the infant, priorities in keeping with a country worried over its demographic future that considered premature infant mortality a national priority. American incubator advocates, on the other hand, were far more infatuated with the device as a symbol of science and modern technology. They were most inspired by the complex incubators of Alexandre Lion rather than the humble *couveuses* of Tarnier and Budin. But if American physicians were so in love with technology, why did they turn against the incubator after 1910?

On one level, the answer involves the paradox that Americans, having invested such high hopes in technology as the answer to the problem of raising a premature infant, rapidly made it a scapegoat when those hopes were shattered. In the French maternities, the incubator never assumed quite the same status as a potential panacea. Tarnier and Budin employed the incubator as an extension of the mother or nurse rather than a replacement; indeed, the glass incubator made possible the coexistence of mother and infant on a cold hospital ward. The wet nurse, rather than the incubator, represented the principle competitor to the mother. American pediatricians attempted for the most part to raise premature infants apart from their mothers. In the infant hospital setting, a formidable gulf of distrust and misunderstanding separated pediatricians from mothers. As a result, premature infants were often doomed by the time they had arrived at the hospital, not because they had been deliberately neglected but because parents were unlikely to consider a hospital as long as their infants were thriving. It was not surprising that incubators failed to function well in this setting. Their rejection, however, reflected the incubator's value as a symbol of the larger problems facing the infant hospital. One party, represented by L. Emmett Holt, remained optimistic about its ability to reform the environment to raise a premature or delicate infant; it merely expanded its focus from the incubator to the hospital ward as the framework for that reform. The other party, typified by Henry Dwight Chapin, rejected both the incubator and the hospital environment as artificial and deleterious to infant life.

The greatest difference between the French and American approaches to the premature infant was not that one was social and the other technological; it was rather that the former integrated the two while the latter divided them. Part of the explanation for this divergence was no doubt the context of French depopulation anxiety, which encouraged physicians to look beyond short-term results in the hospital. But medical specialization, with respect to both professional organization and hospital setting, played a critical and less appreciated role as well. Physicians were prominent in the infant mortality crusades of both countries; in France, however, they tended to be obstetricians, while in the United States they were pediatricians. If infant mortality could be envisioned as an unknown territory between birth, on the left, and childhood, on the right, the two types of physicians entered from opposite directions. French obstetricians retained their primary focus on the mother even as they extended it to the newborn. Their approach centered on the mother but incorporated simple technologies such as the *couveuse*. Pediatricians in the United States had been shaped by a very different experience and were slower to discover the significance of neonatal mortality as well as the benefits of involving the mother. American obstetricians, meanwhile, increasingly centered on the mother alone. In the United States, the premature infant in the early twentieth century was literally caught between two specialties.

CHAPTER EIGHT

The Eclipse of the Incubator

Few American physicians saw any future role for the incubator by the time of the First World War. Although a number of prominent pediatricians and obstetricians had tried to integrate premature infant care into their profession's broader agendas, their efforts ultimately failed. The underlying problem was a fragmentation of neonatal care. Obstetricians concentrated on the incubator and the prevention of early mortality, pediatricians on feeding and the prevention of infection, and neither side had particularly great success. The incubator to a great extent vanished from medical practice during the second decade of the century and would not become popular again until the 1930s.

It would be wrong, however, to see the years 1910–30 simply as an interruption of a chronicle of progress. The flip side of the decline of the incubator was the rise of an aggressive prenatal care campaign centered on maternal education. During these years, in fact, American infant mortality reformers finally discovered the newborn. A sometimes bewildering array of layperson organizations, in which women played a substantial role, abruptly challenged the medical profession's focus on treatment. The principal consequence of this activity was to shift the medical discourse to the prevention of premature birth, a strategy of attack that now appeared far more efficacious than the incubator. But as is often the case with social movements, the impact of the infant mortality campaign on the medical profession was more complex than it appeared at first sight. While principally acting to divert attention from premature infant therapy, in certain contexts it laid the groundwork for its revival in later years. In particular, the new attention on the newborn incited pediatricians to challenge their obstetric colleagues for control of the newborn. The foundations were thereby laid for the rise of a new generation of pediatric newborn specialists, led by the individual who eventually became the country's best-known advocate of the premature infant, Chicago's Julius Hess.

Infant Mortality Reform and the Newborn

The first ten years of the twentieth century gave rise to an extraordinary American crusade directed at the reduction of infant mortality. Although health professionals had been calling attention to the annual "slaughter of the innocents" in eastern cities throughout the late nineteenth century, never before had so much public attention been focused on the problem. In part, the American campaign was but one offshoot of an international movement that had begun in France and spread throughout Europe and Britain.[1] American reformers sometimes spoke in terms of national deterioration and deficiency, recalling the language of European and particularly French reformers. But the context of the Progressive reform movements, which in many respects centered on the child, pervaded the American campaign as well.[2] Like many other crusades of the time, the infant mortality campaign began as a myriad of local organizations joined together to form regional and national networks. At first dominated by physicians, these associations increasingly involved laypersons. Their expansion culminated in the creation of the American Association for the Study and Prevention of Infant Mortality (AASPIM) in 1909, whose bylaws required that physicians account for no more than two-thirds of the elected directorship.[3]

Women furnished much of the leadership and workforce of the growing movement. Drawing on a long-established tradition of forming voluntary organizations outside the home, numerous American women saw the reduction of infant mortality as part of their special domain of "municipal housekeeping."[4] In the nineteenth century, women had been to a great extent responsible for the creation of infant hospitals; the New York Babies' Hospital, which rose to national prominence under L. Emmett Holt, was initially founded by two women physicians and directed by a women's board.[5] As the medical profession consolidated its control within the hospital realm, women led the way in developing alternative approaches to infant health. Public health, and particularly maternal and infant health, offered promising opportunities for women trained in medicine who were excluded by mainstream professional organizations, as well as for nurses seeking independence from medical supervision.[6] Social work offered another possibility; in this case, the settlement movement shaped the values and experiences of numerous future infant welfare leaders.[7]

The infant mortality campaign was part of a broader child welfare movement that brought women from such varied backgrounds together and energized a nationwide network of women's organizations. At New York's Henry Street settlement, Lillian Wald pioneered the city's first visiting nursing service, which was soon seeing more children than were the city's hospitals. Her success later inspired the director of the city's Bureau of Child Hygiene, S. Josephine Baker, to create a similar nursing service to visit all newborns in the entire city.[8] On a national level, no event better symbolized the rising involvement of women in infant welfare than the creation of the Children's

Bureau in 1912. Under the direction of Julia Lathrop, an alumni of Chicago's Hull House settlement and close friend of its leader Jane Addams, the bureau activated on repeated occasions a nationwide network of women's clubs and organizations in carrying out its infant mortality surveys and educational programs.[9]

The settlements also contributed to a shift in emphasis from an environmental to maternal educational focus in the infant mortality campaign by the time of the AASPIM's creation in 1909. To some extent, the newer approach recalled that of Pierre Budin in France, whose *Le Nourrisson* was translated into English at this time and whose example helped inspire the New York Milk Committee to begin providing advice on breast-feeding as well as dispensing milk.[10] But the idea of educating the mother resonated with the settlements' emphasis on educating and assimilating immigrants into American life, as opposed to the hand-out approach they associated with traditional charity. Furthermore, infant mortality statistics of the early century seemed to indicate, possibly as a result of shifting disease categorization, that the clean milk campaigns had failed to reduce infant mortality substantially by the end of the century's first decade.[11] Mounting frustration helped set the stage for the enormous popularity of the British writer George Newman's 1906 book, *Infant Mortality: A Social Problem,* which more than almost any other contemporary work helped to crystallize the new approach.[12] Newman argued that the key to reducing infant mortality lay in maternal education, not in improving the milk supply. Maternal reform quickly became the central organizing principle of the AASPIM and its leaders. New York's S. Josephine Baker made the city's Bureau of Child Hygiene a model for the nation, developing visiting nursing programs to supervise home care and "Little Mothers' Leagues" to instruct older sisters in the care of their infant siblings.[13]

Another rationale for the shift to maternal reform was slower to develop: a recognition of the importance of other causes of infant mortality besides diarrheal disease. The earlier emphasis on preventing gastrointestinal illness by improving the milk supply had overshadowed the issue of neonatal mortality. "The chief cause of infant-deaths is diarrhea, yet prematurity kills half as many," complained the University of Minnesota obstetrician Jennings Litzenberg in 1908. "Does it receive half the attention? I think not."[14] One of the attractions of maternal reform was that it might conceivably affect the mortality of the two causes of infant mortality that vied for second place behind diarrhea—respiratory diseases and neonatal conditions (principally prematurity). But there was much ambivalence regarding the neonatal conditions, since the newborn's constitution had often been regarded as primarily a product of heredity. Such reasoning, if anything, gained further legitimacy as eugenic ideals became increasingly widespread in the early twentieth century. Eugenic reformers took a step beyond the earlier social Darwinist critique that hospitals promoted the survival of the unfit, arguing that society should play an active role in improving the quality of the nation's population.[15] Though infant welfare leaders were, by virtue of their enterprise, hostile to the argument that high infant mortality was nature's way of weeding out the weak and feeble, most accepted more moderate eugenic goals, such as restricting marriage among those presumed to be mentally and constitutionally inferior.[16] This strategy promised to offer more for the newborn than did its treatment afterward.

The emphasis on heredity as a determinant of newborn mortality had particular implications for premature infants, who in this context were seen as products not of accidents but of a diseased maternal environment. Again, the shadowy distinction between premature and congenitally weak newborns reinforced the potential role of heredity. George Newman highlighted the importance of maternal conditions such as syphilis, tuberculosis, alcoholism, and fatigue as causes of premature birth.[17] The same conditions that caused premature death could also cause atrophy and wasting in the survivors—a condition he described as "immaturity." In his view, the outlook for such infants was guarded at best. "Infants thus heavily handicapped may yet survive for some weeks and even grow up to adult life," he wrote. "But the mark of death is upon them."[18]

Yet by the time of the First World War, another fruit of the rising national concern over infant mortality, the improvement of vital statistics, catapulted the issue of neonatal mortality to public attention, in spite of eugenic reservations. The 1910 census, whose results were first published in 1913, marked a particularly significant turning point in this regard. Unlike earlier vital statistics, this was the first census in the United States to provide infant death-rates by age, according to the number of days, weeks, and months of life. Its figures vividly dramatized the law of nature espoused by nineteenth-century sanitarians, that mortality increased with younger age. Of the registered infant deaths in 1910, 38 percent took place within a month after birth. Causes unique to early life, particularly prematurity and its ill-defined cousin, congenital debility, contributed 61 percent of deaths in this period.[19] Widely discussed in infant mortality circles, the new statistics played a major role in bringing the newborn to the forefront of the AASPIM's 1914 annual meeting.[20]

The "discovery" of the newborn by infant mortality reformers set into motion two powerful but conflicting forces with respect to the premature infant. On one level, it provided physicians a new incentive to address its mortality; as had been the case in France, the treatment of the premature infant might now become a public health imperative rather than a medical challenge. The potential arose for new allies and new sources of support, along with the possibility that premature infant advocates might no longer have to work in isolation, as had Rotch and DeLee. At the same time, the eugenic assumptions underlying the infant mortality movement did not vanish following the discovery of the newborn. The newly awakened interest of infant welfare reformers in the newborn raised possibilities for conflict as well as cooperation.

A Maternal Approach

The first and most significant consequence of the public health community's new focus on neonatal mortality was the rise of an organized prenatal care movement. Although the causes of prematurity remained obscure, few questioned the basic proposition that a pregnant woman leading a healthy and "natural" lifestyle would be less likely to deliver her infant prematurely. Reformers assumed that the elements of hygienic living (plenty of wholesome food, fresh air, and the proper balance of rest and exercise) would offer protection against both miscarriage and premature birth.[21] The New York Milk Committee (NYMC) and the Boston Lying-in Hospital had begun advising pregnant women on hygiene as early as 1906. Stimulated in part by the interest generated by the Russell Sage Foundation's influential 1910 study of European prenatal care, the promoters of these programs conducted statistical studies of their own during the next several years.[22] The NYMC, for example, had sent visiting nurses at regular intervals to provide education and referral to some 1,375 pregnant women by 1913; it subsequently reported a neonatal death rate 32 percent lower than that of Manhattan in general.[23] In Boston, the Women's Municipal League conducted a similar program that also attained premature and neonatal mortality rates less than half that of the city at large.[24] Though these studies could be faulted for using city statistics as controls, they generated great excitement on behalf of prenatal education.[25] Maternal education and supervision, it appeared, could greatly reduce neonatal mortality with only rare recourse to the services of a physician.

Beyond its apparent efficacy, the great appeal of the prenatal care strategy was that it avoided the troubling eugenic objections to the actual treatment of premature newborns. For example, it is striking that Mary Mills West, the author of the enormously popular Children's Bureau booklets *Prenatal Care* (1913) and *Infant Care* (1914), confined all specific discussion of premature birth to the former. While offering considerable advice on the importance of preventing early labor through such hygienic measures as a laxative diet, two hours daily in open air, and the avoidance of alcohol, West said nothing about the care of a premature infant born despite such measures. Indeed, she recommended a room temperature of 65°F even for very young or delicate babies.[26] West's decision to focus exclusively on prevention was hardly a matter of oversight, for, judging from a 1915 address to the AASPIM on prenatal care, she had great reservations regarding the future of surviving premature infants: "These puny, ill-conditioned babies crowd our welfare stations and hospitals; many of them die in later infancy. . . . Others live to recruit our institutions for sick, crippled, deficient, and defective children; still others live on dragging out enfeebled existences, possibly becoming finally the progenitors of weaklings like themselves."[27] Prenatal care advocates, in fact, often voiced concern that surviving premature babies would be relegated to a "permanently enfeebled life."[28] Prevention appeared to provide a way to sidestep the difficult ethical issues associated with therapeutic care. As summarized by West, "Prenatal care will tend to the production of a race of vigorous, normal babies."[29]

Such an outlook hardly encouraged the hospital treatment of premature newborns. Indeed, as West's testimony implied, many infant welfare advocates were inclined to see the hospitalized premature infant through the same lens in which they viewed the institutionalized atrophic infant. The premature baby remained a weakling for much of the public and, to some extent, for public health reformers as well. This perspective had considerable significance, for the half-century-long dispute over infant and child institutionalization was coming to a head by 1915. The opening salvos had been fired by Lillian Wald and her visiting nurse allies, who increasingly defended their success in terms not merely of individualized care but of the intangible influence of the mother.[30]

The treatment of hospitalism moved from the level of the physical environment to that of the psychological. One of the key figures in this regard was none other than Henry Chapin, whose retreat from the incubator had now expanded into a broader assault on the infant hospital. Although Chapin had begun his experimentation with supervised home medical care (through the Speedwell Society) in the name of a better environment, he eventually became convinced that its success depended on more than fresh air or efficient care. By 1911, he joined the critics of pediatric expertise and the infant hospital; infants, he wrote in a style hardly likely to cool passions, required "the affection that even an ignorant but well meaning woman with motherly instincts can often give better than a nurse with the highest theoretical training even under a doctor in whom abstract science may be higher developed than common sense."[31] In an influential 1914 paper titled "Are Institutions for Infants Necessary?" Chapin decried before the AASPIM the proliferation of infant hospitals around the country.[32] In doing so he moved well beyond the pediatric mainstream, marking himself as a radical who no longer saw social pediatrics as compatible with the academic style of his colleagues.

It was in this context that S. Josephine Baker, the fiercely independent and innovative director of the child hygiene division of New York's health department, explicitly drew premature infants into the institutionalization controversy. Encouraged by her success with earlier maternal supervision programs, she developed a boarding-out program for the premature, atrophic, and marasmic infants consigned to the "hopeless ward" of the New York Foundling Hospital, as they had been in the nineteenth century. Her description suggested that little had changed, despite the introduction of incubators and modern hygienic technique, for the death rate on that ward was essentially 100 percent.[33] "Any scrawny, bluish, half-alive baby that went there, to be wrapped in cotton wool and fed with a medicine dropper," she later recalled, "was morally certain to fade out after a little while, and through nobody's fault at all."[34] Baker induced the Russell Sage Foundation to provide an additional five dollars a month to pay foster mothers to take these fragile patients into their care under the supervision of physicians and nurses. The infants, most of whom had been in incubators, were in such precarious condition that nurses were required to supervise their transportation from the sterile and orderly wards of the hospital to the tenements of Hester and Orchard Streets.

The concept of the premature infant transport system had been turned on its head. Yet Baker found that in spite of the tenement environment, the foster mothers reduced the mortality rate of the infants from nearly 100 percent to just over 50 percent, at a cost less than half that of hospital treatment.[35] In 1915, one year after Chapin's attack on the infant institution at the AASPIM, she presented the study to the same forum as part of her own contribution to the institutional mortality debate. Baker interpreted the results as a testimonial to not merely the supervision of physicians and nurses but "possibly most important, the human element.... These babies are very delicate and very ill; devoted love and the deepest human interest must be given in abundance."[36]

The premature infant, once a symbol of the potential of science, now provided a tool with which to attack the infant hospital. Yet the attack centered not on technology but on the mechanical nature of institutional life, the growing impersonality and efficiency characterizing the hospital. And it conceptualized the premature infant as a weakling requiring individualized care and love rather than scientific expertise alone. "The efforts of medical science may pertain equally in the institution or in the home," Baker concluded, "but the vital need of every baby to be considered as one human being and not one of a group must take precedence over all other considerations."[37] Academic pediatricians were quick to dissent. Alfred Hess, one of the most brilliant of his generation of young pediatric investigators, challenged the conclusions of Chapin and Baker while delivering his own paper to the AASPIM.[38] Arguing that hospitalism was not "a pervading evil genius" afflicting the wards but simply the phenomenon of frequent infection resulting from inadequate facilities, nursing, and medical attention, Hess insisted that babies on the wards not be handled more than absolutely necessary. "In this respect," he continued, "we differ absolutely from those who believe that the 'mothering' of infants is necessary. . . . On the contrary, this handling or mothering spreads the grippe infection just as it disseminates measles or other communicable diseases."[39]

Baker's appeal to the importance of "mothering" in some ways recalled Budin's maternal approach in France, with one substantial difference. Like Budin, she conceptualized premature birth as a social rather than a purely medical problem, one that involved economic as well as educational issues.[40] Yet Baker, as a female infant health physician, had a different perspective from the male obstetrician's. In this regard she represented one of the most successful cases of an American woman physician who responded to the exclusivity of male-dominated specialty organizations such as the APS by entering public health. There was a feminist message behind Baker's insistence on the importance of mothering, one that reasserted the idea that women possessed a special intuition for caring for a sick infant.[41] This is not to say that maternal reform served a fundamentally ideological purpose in justifying the involvement of women (physicians, nurses, and social workers) in supervising infant care. Many male physicians took up the banner as well, and Baker did not see herself as an ardent feminist, in spite of her support of women's suffrage. Nonetheless, it does seem fair to suggest that Baker's maternal approach exhibited a greater degree of sympathy for mothers than that expressed by Budin. In particular, her rhetoric did not stress, as that of the French obstetricians so often did, the bearing and raising of children as a duty.[42]

Indeed, on the issue of providing economic as well as educational assistance to the mother, Baker went beyond many infant mortality reformers as well as physicians. It was quite possibly this issue, more than any other, that eventually undermined the potential of the Progressive prenatal care movement. As historian Richard Meckel has pointed out, the American infant welfare community shared with its European counterparts a common focus on the mother but parted ways with respect to whether maternal reform included economic support as well as education.[43] By 1910, a number of European nations, first Germany and later France and Britain, had developed maternity insurance programs to provide material assistance to women in pregnancy and the postpartum period. Such measures attracted considerable attention as part of broader sickness insurance proposals in the United

States during the First World War but, ultimately, failed to win passage in a single state. Many factors were clearly involved; the United States had little infrastructure to administer such measures, its medical profession (after some initial flirtation) came to oppose government insurance; and the German origins of state-supported medical insurance made it suspect once the United States entered the war.[44] Yet most infant welfare leaders were ambivalent, at best, toward offering any financial support to induce women to fulfill what were supposed to be their moral obligations to their children.

Ironically, even the Children's Bureau, which conducted studies providing powerful support in favor of an economic interpretation of neonatal mortality, took little part in actively promoting economic assistance for mothers. Between 1915 and 1923, it conducted a series of ten community studies linking infant mortality to economic status.[45] In particular, the surveys pointed out that women with little family income were far less likely to seek prenatal care. Such findings raised concern over whether women of limited means could carry out the standard prenatal regimen of freedom from overwork, two hours daily in open air, and a nourishing and wholesome diet.[46] The numerous letters sent by mothers to the bureau requesting advice made the shortcomings of maternal education still more glaring.[47] One woman wrote Julia Lathrop, the bureau's director, in 1916,

> Will you be so kind and send me advice [on] how to gain a little more strength. . . . I am so run down I can hardly do my own work as I put all my time on baby, but if I were stronger I could do a lot more for him and myself. I am 22 years old and 5 ft. 3 in. and only weigh 96 lbs. I am now again an expectant mother of three months past, and am so afraid my child will not be strong and normal on account of my condition.[48]

Many such letters sounded a common theme. A Chicago woman wrote at length in the same year of how her own circumstances made the ideal of "rest" during pregnancy nonsensical and concluded: "Now, sirs, I am to become a mother again this coming month . . . and nothing to nourish the coming. only *abuse* and *torture* at the hands of the man who *promised* to *provide* and *protect woman* [*sic*]."[49] Typifying the personal care with which Children's Bureau members often answered such inquiries, Lathrop sent a copy of this particular letter to her physician friend Alice Hamilton of Hull House, imploring her to "see if there is some human way to help her. . . . This type of letter is becoming too frequent."[50] Yet while Lathrop was clearly aware of the importance of economic circumstances to prenatal care, she never played a major role in translating this concern into support for maternity insurance. A pragmatic leader by instinct, she increasingly deemed it necessary to concentrate her energy on less controversial educational measures as the United

States approached the more conservative 1920s. Her crowning achievement promoting maternal and infant welfare, the Sheppard-Towner Act of 1921, thus embodied the educational approach to prenatal and infant care.[51]

Although the prenatal care movement that originated in the Progressive Era and reached its high point in the 1920s certainly deserves more study, we must limit our own examination to those aspects directly relevant to premature infant care. Perhaps the main point to underline is the extent to which education in the principles of prenatal hygiene appealed to many infant welfare reformers as almost a panacea for neonatal mortality. The extraordinary optimism that followed the first studies of prenatal care understandably diverted attention from the treatment to the prevention of prematurity. Such hopes eventually proved difficult to sustain; the potential of prenatal care was undermined by both medical ignorance of the causes of premature birth and the failure to back educational advice with material support. Yet few appreciated these limitations until the 1930s, when the pendulum would again swing back toward treatment.[52]

Yet while the main thrust of the infant mortality crusade was to shunt energy away from the treatment of premature newborns, its effects were not completely negative. By highlighting the importance of newborn mortality, it raised professional consciousness of the extent to which the newborn had been overlooked in previous years. This heightened awareness did not translate into a resurrection of the incubator, which was already discredited by this time. But it did generate a certain degree of interest in both examining the newborn anew and redefining the problems it posed. In the process, obstetricians and pediatricians negotiated once again their boundaries with one another as well as their responsibility for the newborn.

No-Man's-Land

The new recognition of neonatal mortality brought obstetricians to the forefront of the infant welfare movement. The election of an obstetrician, John Whitridge Williams, to the presidency of the AASPIM in 1914 defined, as much as any single event, the transformation. Williams has been described, along with DeLee, as one of the two titans of twentieth-century American obstetrics. He did in fact write the only textbook in his specialty whose popularity directly rivaled that of DeLee and, as the first professor of obstetrics at Johns Hopkins, trained many of the next generation's most influential leaders. Yet his personal background and style could hardly have differed more from DeLee's. The descendent of long-established medical families on both sides, Williams radiated a sense of ease and grace matching his genteel background and cultivated a professional life where a case a week was regarded as ample.[53] His therapeutic style was also more conservative than

DeLee's, manifested by a less enthusiastic response to cesarean section in his early years and outright opposition to the prophylactic forceps approach in the 1920s.[54]

A similar conservatism marked his attitude toward the treatment of premature and weak newborns, about whom he was far less optimistic than his Chicago colleague. This orientation had already led him to respond favorably to prenatal care after meeting the director of the innovative program devised by the Boston Women's Municipal League.[55] In an address at the opening session of the AASPIM in 1911, he had argued that prenatal care was far more important than treatment during the first weeks of life. "We must also learn that the birth of a puny or damaged child is a great misfortune," he asserted, "as its chances of reaching maturity are greatly diminished, not to speak of the additional expense to the individual or the state entailed in rearing it."[56] Williams expressed a sentiment that events would soon reveal to be widely shared by his obstetric colleagues.

Williams's presidential address to the AASPIM on "The Limits and Possibilities of Prenatal Care" outlined a new, and explicitly biomedical, approach to the reduction of neonatal mortality. Specifically, he underscored the relationship between congenital debility and congenital syphilis. Physicians, as we have seen, had been interested in the role of syphilis as a cause of premature birth since the nineteenth century and had often asserted that the disease could induce a state of hereditary weakness in the afflicted offspring. Much of this testimony was admittedly speculative and based on a far more fluid concept of the disease's transmission than that accepted today. Nonetheless, a series of dramatic breakthroughs in the first decade of the new century reenergized the entire question of the connection between syphilis and prematurity. Foremost among these were the successive discoveries of the treponemal spirochete causing the disease (1905), the Wasserman blood test that facilitated its diagnosis in (1906), and the chemical "magic bullet" Salversan that promised cure (1909).[57]

Williams reported the results of a study of the incidence of the Wasserman reaction in a series of ten thousand consecutive births at the Johns Hopkins obstetric wards. He centered his analysis on 705 "foetal deaths," newborns of at least seven months' gestation ("the so-called period of viability") who died either upon birth or during their two-week routine hospitalization. The series made no attempt to distinguish stillbirths from early postnatal deaths, a point to which we shall return. Williams reported that nearly half of the fetal deaths (334 of 705) were premature by his definition (less than twenty-five hundred grams). But his more sensational finding had to do with the results of routine Wasserman screening on placental blood. A quarter of the tests were positive, leading Williams to conclude that in his study population, "syphilis is far and away the most common etiological factor in the production of death." Indeed, the rate was still higher for premature infants, accounting for 40 percent of all deaths associated with prematurity and a still higher 49 percent of deaths among those who were black as well.[58] Widely quoted by physicians and infant welfare reformers during the following ten years, Williams' study not only strengthened the case for prenatal care as an alternative to treatment but also rationalized a role for the medical profession and laboratory medicine in prenatal care.

Indeed, Williams's paper epitomized the biomedical orientation that characterized the specialty's approach to prenatal care during the following decade. To be sure, Williams' point was not to reject the utility of maternal education as conducted by visiting nursing services and infant welfare stations but to define an appropriate role for his own profession in the wider campaign.[59] Nonetheless, obstetric leadership did have the effect of distracting attention away from the question of economic versus educational maternal reform and toward that of educational versus medical intervention. It also highlighted issues relevant to professional agendas in other respects. For example, obstetricians tended to concentrate on the prevention of birth injuries, even though they accounted for only a small percentage of neonatal mortality. Because injuries were rooted in the management of labor, they were relevant to the concurrent battle between obstetricians and midwives.[60] The value of incorporating syphilis screening into prenatal care, on the other hand, had more to do with raising the status of obstetrics itself along the lines of laboratory medicine.

Unfortunately, Salversan no more fulfilled its promise as a "magic bullet" for syphilis in the newborn than it did for the adult.[61] Williams attempted to screen and treat all maternity patients seen at the Johns Hopkins clinics during the following five years after his 1914 address but had great difficulty in assuring adequate treatment. By 1920, he was considerably more circumspect about the intervention, noting that he had been able to provide a full course of four to six Salversan injections to only 163 of his 421 mothers with positive Wasserman tests.[62] Moreover, the maternal Wasserman turned out to be an imperfect predictor of congenital infection; less than half of the babies born to mothers with a positive test were syphilitic, while in rare cases mothers with negative tests gave birth to afflicted infants.[63] Williams eventually conceded that his strategy offered no panacea for the prevention of deaths from prematurity, whose causes he believed were for the most part unknown.[64]

While obstetricians tried to analyze the mortality of the newborn period from the standpoint of its prevention through prenatal care, pediatricians became conscious of their own previous neglect of the neonate as well. Indeed, they found themselves somewhat on the defensive, as their obstetric colleagues suddenly threatened to outflank them as the leading medical professionals within the infant welfare movement. One of the first to respond was L. Emmett Holt, who presented at the 1914 AASPIM meeting a counterpart paper to Williams's presidential address. In conjunction with the nurse and infant welfare reformer Ellen Babbitt, Holt conducted his own study of mortality during the first two weeks of life among ten thousand consecutive births at the Sloane Hospital for Women. His method differed from Williams's in that he excluded 429 stillbirths and limited his analysis to the live-born babies, of whom 291 had died in the first two weeks of life.[65] Birth accidents accounted for only thirty-three of these deaths, "an astonishingly small fraction of infant mortality" in Holt's estimation.[66] Holt also found few patients with congenital syphilis, although he readily acknowledged that Sloane did not accept known syphilitic mothers and that, therefore, few Wassermanns were done.[67]

In his study population, prematurity without any accompanying condition other than "congenital weakness" accounted for half of all deaths in the first two weeks. "This, then, is one of the largest and most important factors of infant mortality," he concluded. It was also one the obstetricians seemed to have forgotten, as Holt implied by adding that its cause lay in the mother's physical condition rather than the circumstances of her delivery.[68] Holt's influential study illuminated the mortality of the newborn within the hospital just as the 1910 census had done for the country at large. It also showed how different the pediatric and obstetric perspectives on this mortality could be.

Of particular interest is how Holt and Williams diverged in drawing the line between premature infants and stillbirths. In counting together as "foetal deaths" all infants born prematurely, regardless of whether they had been born dead or had died in the nursery thereafter, Williams was conducting a practice condemned by virtually all leading pediatricians and infant mortality workers. At the first meeting of the AASPIM in 1910, Abraham Jacobi had deplored the variable definition of stillbirth in many communities. "Deaths of infants 2 weeks old or under *may be*, and in some cities *are*, thrown out of the mortality count," he asserted.[69] Holt and Babbitt made a similar point in a less polemic fashion in their 1914 report. They believed stillbirths to be the most unreliable part of all vital statistics, one that, depending on the locale, might include infants who died within two hours, two days, or even one week.[70] Williams's practice, on the other hand, recalled the tendency of nineteenth-century obstetricians to see prematurity through the lens of "apparent death" and asphyxia, an attitude that combined fascination with the possibilities of acute resuscitation with fatalism regarding infants who responded but remained weak thereafter. His classification, moreover, had considerable medical importance in highlighting the role of syphilis in prematurity, since congenital syphilis is associated with a high rate of stillbirth as well as prematurity.[71] Conversely, to the extent that pediatricians tightened the distinction between stillborn and live-born infants, they were less likely to emphasize the connection between syphilis and prematurity.

Holt's study marked one of the early signs of a pediatric advance into the obstetric nursery. This heightened interest in the newborn took place in response not only to new statistical studies but also to a phenomenon in the second decade of the century with great future implications: the entry of middle-class women into the maternity ward. By the time of Holt's study, hospitalized childbirth was no longer restricted to the poor; Sloane Hospital, in fact, had been accepting a large percentage of private patients since the late 1890s.[72] The years between 1910 and 1920 have often been regarded as a turning point in this longer process, at least in larger cities. By the time of a Children's Bureau survey in 1922, over half of all women in some cities such as Minneapolis and Washington, D.C., were delivering their infants in the hospital.[73] The causes of this momentous shift in attitude have been themselves the objects of considerable historical scholarship, with more recent interpretations underlining the role played by women themselves in seeking safer and more pain-free labor by entering the hospital.[74] Pediatricians were initially bystanders in this process; largely unaware themselves of the significance of neonatal mortality, they hardly saw its reduction through hospitalized childbirth as a high priority. But as larger numbers of mothers and thus newborns entered the hospital, pediatricians soon became aware of a potential field for expansion. In doing so they formally challenged the traditional obstetric responsibility for the newborn.

In the wake of the 1910 census results, pediatricians thus began to organize a campaign to gain access to the obstetric nurseries. The leaders of the profession shot off the opening salvos. At the dedication of the Harriet Lane Home, the new children's hospital at Johns Hopkins, the normally circumspect Holt remarked with uncharacteristic acerbity how "in maternity hospitals infants are tolerated as one of the unavoidable incidents of obstetric practice."[75] Abraham Jacobi delivered a particularly stinging comparison of obstetricians with their professional nemesis by remarking in his 1912 presidential address to the American Medical Association that many premature and feeble infants could be saved with ample care, "such as a midwife will more readily give than a doctor."[76]

But it was the younger members of the profession, intent on finding new career opportunities, who provided the foot soldiers for the battles to follow. In 1916, the Association of American Teachers of the Diseases of Children (AATDC) initiated a formal effort to bring newborn nurseries under the control of pediatricians. This organization had emerged in 1907 out of the AMA Section of the Diseases of Children and became a second professional pediatric organization whose membership in the South and Midwest contrasted with the northeastern, elite medical school orientation of the American Pediatric Society (APS). By this time, the AATDC was directing the mobilization of pediatrics as an autonomous academic specialty within medical schools far more aggressively than had its older cousin, the APS.[77] Its drive to take charge of the nursery fit neatly into a broader program of explicitly defining the boundaries of pediatrics against both medicine and obstetrics.

The advance of pediatrics into the obstetric nursery nonetheless soon deteriorated into trench warfare in the face of persistent opposition and inertia in many institutions. Progress took place slowly and unevenly, as the responses of obstetricians varied among the extremes of hostility, apathy, and cooperation. Some resisted pediatric involvement out of genuine concern for the newborn, in the belief that breast-feeding required the obstetrician to supervise both mother and infant.[78] Indeed, we should be careful not to accept assertions of obstetric "neglect" made by pediatricians at face value, given the professional incentives to perpetuate such a portrayal. Even allowing for this qualification, however, there was probably considerable truth behind the pediatric charges. Obstetricians, tellingly, only rarely wrote articles on the newborn, much less published rejoinders to the pediatric allegations. No leading American obstetrician by the time of the First World War showed as much interest in the newborn as did Pierre Budin in France or J. W. Ballantyne of Britain. Ballantyne, interestingly, became one of the first physicians to borrow a term from the Western Front to describe the two specialties' stalemate over the newborn. In a 1916 address, he described the newborn infant as occupying a "No-Man's Land" between obstetrics and pediatrics.[79] The term caught on and, indeed, remained in fashion throughout the 1920s.[80]

It is hard to trace the exact progress of this struggle, since medical journals featured more rhetoric and exhortation than assessment of actual practice. But we should certainly be cautious in assuming any neat correlation between pediatric control of the nurseries and the national reputation of the institution involved. For example, the pediatric department formed in 1911 under John Howland at the Johns Hopkins Medical School soon acquired a national reputation as perhaps the most distinguished in the country. Yet Hopkins was conspicuous mainly for its absence with regard to the early history of newborn care. The pediatricians were based in a children's hospital (the Harriet Lane Home), while obstetricians controlled the nursery at the main hospital.[81] J. Whitridge Williams, as already noted, viewed efforts to treat (as opposed to prevent) prematurity with suspicion. Like many contemporaries, he manifested considerable eugenic inclinations, and in one address to the AASPIM he went so far as to suggest that it would be more humane to legalize the destruction of institutionalized infants than leave them to a fate of slow starvation without individualized care.[82]

The Johns Hopkins experience may well have pointed to the future of premature infant care in another respect, however: the role played by race in shaping the perception of prematurity. Reflecting its geographical situation, Hopkins was unusual among major East Coast academic centers in admitting a substantial percentage of black patients. Williams was one of the first physicians to document that their neonatal mortality rate was nearly twice that of white infants. Syphilis was over twice as common as well, suggesting the desperate social circumstances from which many must have emerged.[83] Williams attributed many of the deaths from prematurity to his hospital population. Addressing the prevention of premature birth as part of his 1914 AASPIM address, he asserted that "cases of this character occurred much more frequently among the blacks and are attributable, in part at least, to the lack of care and intelligence which so frequently characterizes that race."[84] The issue of race, and of the higher rate of premature births found among black patients in particular, would play an increasingly prominent role in discussions of prematurity in later years.[85]

At least in some hospitals, obstetric practice continued an unwritten policy of benign neglect of premature and congenitally diseased infants. This "policy," to be sure, was rarely articulated; instead, it was implicit within the language used to characterize sick newborns. In this respect, the willingness of obstetricians to use the term *stillbirth* in a flexible sense to cover early neonatal deaths conceivably had great relevance. One might speculate that the term served a useful social purpose in rationalizing decisions not to provide systematic treatment to small premature infants. Evidence for this assertion generally comes from later years, when pediatricians were becoming more aggressive in challenging obstetric practice in the nursery. For example, in 1929 an investigation of Boston's lying-in hospitals (considered among the country's best) found that on its obstetric services, many newborns died within a few hours of birth without any comment of unsatisfactory condition in the medical record. Prematurity was often given as a reason for not examining a baby who should have been expected to survive on basis of its gestation.[86] According to the recollection of a later leader of neonatology, obstetricians continued to write off many small premature infant deaths as stillbirths well into the 1940s.[87]

In other institutions, the transition from obstetric to pediatric responsibility for the newborn proceeded more smoothly. Some obstetricians became pediatric allies, finding the new arrangements advantageous to both sides. Perhaps the foremost obstetric advocate of incorporating pediatricians into the nursery was Jennings Litzenberg, chairman of clinical obstetrics at the University of Minnesota. His own interest in the newborn having been sparked in part by the example of Joseph DeLee's incubator station in Chicago, Litzenberg defended the pediatric cause before both professional and public health forums.[88] Litzenberg's attitude derived from his own experience of granting control of the nursery to the pediatric department; although he had initially expressed misgivings, he soon became convinced that the newborns received better care, while the obstetricians had more time to focus on the mother. The University of Minnesota pediatric department, in fact, became a leading center of neonatal medicine, with Litzenberg its most fervent obstetric ally.[89] In a paper before a joint session of obstetricians and pediatricians at the AASPIM in 1918, he testified how the surgical and medical perspectives of the two specialties set them apart. "The care of the new-born is neglected in most maternities," Litzenberg asserted, "not because the obstetrician wilfully ignores the baby, but because his habits of thought, training, and desires lead his mind in another direction."[90] With more insight than many of his contemporaries, Litzenberg had nicely summed up the essential problem the newborn posed for his profession: it did not fit into the paradigm of acute care, surgical obstetrics.

On another level, the Minnesota experience pointed to the rise of the West, and particularly the Midwest, in pediatric neonatal medicine. There had already been hints of this development; John Zahorsky of St. Louis had produced the most extensive (if unappreciated) American treatise on the premature infant, and Joseph DeLee of Chicago had created the most advanced incubator station of any hospital in the country. As incubator rooms emerged toward the end of the second decade of the century, New York increasingly gave way to St. Louis, San Francisco, and Chicago as the cities leading the way in developing the new technology.[91] The rise of interest in premature infant care in these cities arose in parallel with the expansion of the pediatric profession beyond the East Coast, symbolized by the rise of the AATDC. But the most decisive indication of the shift in leadership from the East to the West took place in Chicago, where the pediatrician Julius Hess would emerge as the nation's acknowledged leader of neonatal care.

Julius Hess: Integration and Consolidation

Though a full consideration of the career of Julius Hess belongs to the later history of neonatology, a brief consideration of his early work illustrates his continuity with the time period examined in the present study. Hess brought together, to an unprecedented extent, many of the various strands characterizing the early phase of American neonatology. Born in 1876, not long before the incubator's invention, he graduated from Northwestern Medical

School in 1899 and subsequently undertook postgraduate work at Johns Hopkins and later in Germany. After returning in 1901, he set up practice and gradually established himself as one of Chicago's most prominent early pediatricians.[92] At just what point he began to focus on premature infants is uncertain. Like Thomas Rotch and the Boston pediatricians, Hess may well have moved in this direction as a part of a broader interest in infant feeding. His first paper on premature infants, published in 1911, was a study of their caloric requirements and, indeed, was the first such American effort since the studies of Morse and Zahorsky several years earlier.[93] The incubator show tradition seems to have influenced Hess as well, given the intriguing (but conflicting) stories of his friendship with Martin Couney. Most of these accounts place the first meeting between the two men in 1914, but at least one pushed it back to 1905.[94]

It is hard not to conclude that DeLee's presence in Chicago had an impact on the young Hess as well. The surest connection in this regard would seem to be his fellow pediatrician Isaac Abt, who had supervised DeLee's incubator station and in the early 1900s oversaw Hess as head of the Department of Children's Diseases at Michael Reese Hospital.[95] At any rate, Hess probably did have closer ties to other obstetricians than did some of his East Coast pediatric colleagues based exclusively in children's hospitals. Neither Hess nor Abt had spent their early professional career in an institution devoted exclusively to infants and children; both were affiliated with Michael Reese, a general hospital that by 1907 was using incubators in the more advantageous environment of the maternity ward nursery.[96] Hess and Abt, in other words, were among the early wave of pediatricians intent on establishing a role for their discipline in the obstetric nursery.

Another factor linking DeLee and Hess that deserves consideration was religious. It is intriguing that the first two American promoters of incubator stations were Jewish. And both received much of their support from the Chicago Jewish community. Michael Reese was a remarkable institution, established by members of the cultured and predominately German Jewish immigration that had arrived in Chicago in the 1840s. Its patients derived largely from the far less privileged Polish and Eastern European Jewish community that arrived later in the century. Considerable tension separated the two groups, a rift that one suspects may have colored the relations of DeLee and Hess.[97] But it is also possible that the shared experience of discrimination could have led Jewish physicians toward clinical areas neglected by the professional mainstream. And the rhetoric of eugenics, with its nativist overtones, likely had little attraction for DeLee and Hess.[98]

Hess's first original contribution to premature infant care was his invention of a new incubator in 1914. More precisely, he created what he point-

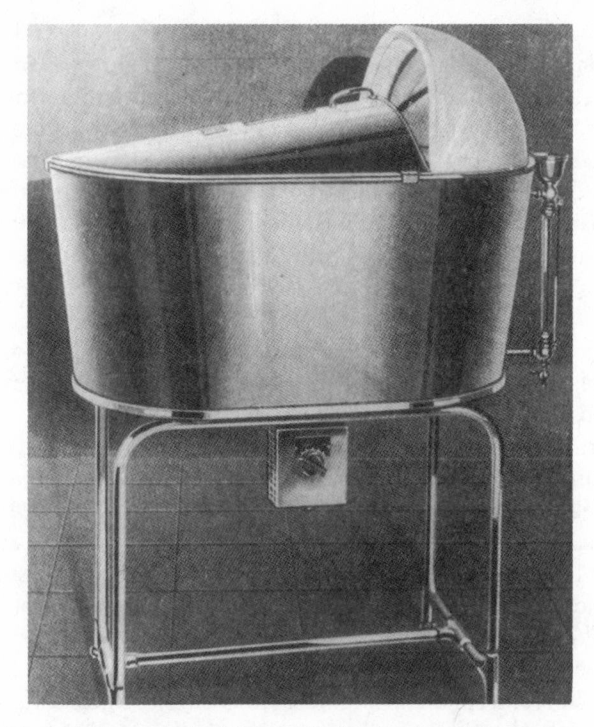

Hess incubator bed. *Source:* Julius Hess and Evelyn Lundeen, *The Premature Infant: Its Medical and Nursing Care* (Philadelphia: J. B. Lippincott, 1941), 36.

edly called an electrically heated "incubator bed" intended to be an alternative to the discredited incubator. The device was essentially an electric version of Crede's *warmwänne,* a double-jacketed tub housing a layer of warm water whose temperature was controlled by a rheostat. In contrast with the traditional enclosed incubator, it featured an open hood allowing air to circulate naturally around the infant. Hess's heated bed thus offered the open air afforded by a padded basket without the inconvenience of frequent hot water bottle changes.[99] It also maintained the intrauterine metaphor of the incubator as a refuge from the outside world. Minimal excitement and stimulation were to become hallmarks of Hess's approach to premature infants.[100]

Although Hess's heated bed fit with American more than with French notions of the incubator, in one important respect it represented a departure from the former. Its precise mechanism for adjusting heat, the electric rheostat, embodied the way in which Hess stood apart from many of his contemporary pediatricians in recognizing the importance of temperature. He rejected incubator rooms primarily because their design implied one temperature to be sufficient for all premature infants; in contrast, Hess insisted that temperature had to be matched precisely to the individual infant.[101] The baby's own temperature, which he insisted should be measured every six hours and recorded on a graph, guided that of the incubator. Hess also displayed unusual awareness of the importance of preserving temperature immediately after birth, another insight that previously had been more characteristic of obstetricians than of pediatricians.[102] By 1922, he became the first American physician since DeLee and Zahorsky to develop a transport incubator. The device was perhaps the most tangible evidence of DeLee's influence on Hess.[103]

Hess's incubator may have appeared somewhat primitive compared with the earlier Lion model, but its creator nonetheless succeeded in building the first permanent American incubator station around it. The critical factor that enabled him to accomplish what had eluded Joseph DeLee was, interestingly, his alliance with a women's voluntary association. In 1914, Hortense Shoen Joseph (interestingly, a prominent member of Chicago's Jewish community and the wife of a wealthy businessman) created the Infants' Aid Society, an organization specifically directed at improving the care of premature infants. Distressingly little is known about the origins of this intriguing association, or why it, seemingly alone among other women's volunteer groups in the United States at the time, chose its particular focus. Although a personal friendship between the Joseph and Hess families probably accounted for part of the explanation, the creation of the society in 1914, just as infant welfare reformers nationwide were discovering the newborn, can hardly have been coincidental.

Joseph and her organization soon formed an alliance with Hess, allowing him to introduce his heated beds into Michael Reese and the other hospitals he was affiliated with.[104] His facilities remained taxed to their limits, however, by patient demand. The Spanish flu pandemic of 1918–19 left in its wake an especially poignant tragedy that dramatized the need for better hospital newborn care. Hess remarked in 1920 that he had never seen so many premature babies admitted after the deaths of their mothers.[105] A second tragedy soon brought about the incubator station he had long sought. Joseph herself died unexpectedly at the age of thirty-two in 1922, leaving an endowment of over sixty-five thousand dollars to create a new incubator station.[106]

The year 1922 marked a turning point in Hess's career and, indeed, in the entire history of American neonatology. Two events were particularly significant. First, with the support provided by the Infants' Aid Society, Hess expanded his battery of incubators into the most sophisticated premature infant nursery of its day. He continued to call the facility an incubator station, again showing his debt to DeLee. Located across the street from Michael Reese at its more recently opened Sarah Lawrence Children's Hospital, the station eventually encompassed twenty-eight incubators. It emulated DeLee's transport system, now assisted by the automobile. Hess also set up quarters for five to eight wet nurses, far more than any previous pediatrician had supported, whom he oversaw with a paternalism worthy of any of the French obstetricians.[107] The station provided Hess with a degree of practical experience with premature infants unmatched in the United States.

Exemplifying the rising authority of the forty-six-year-old pediatrician was the year's second event in this regard, the publication of Hess's textbook, *Premature and Congenitally Diseased Infants.* The title made explicit the conceptual advance represented by the creation of the incubator station: the distinctiveness of the premature from the weak infant. Hess had created a domain within the hospital that separated premature from other weak and feeble infants. He discarded the term congenital debility, affirming the difference between babies suffering from simple prematurity and those with complicating specific diseases.[108] Characteristically, much of his later research would address the long-term outcome for premature infants, both physically and intellectually, to counter the charges that he was salvaging constitutional weaklings.[109] He became the foil to the dominant eugenic philosophy characterizing his era.

These accomplishments won Hess a place in the annals of American neonatology as the father of the specialty.[110] The honor, of course, is most apparent in retrospect. Julius Hess was in many ways out of phase with the medical profession during the interval between 1910 and 1930, a period that marked a hiatus in the history of premature infant care in the United States. These years separated two eras of relatively intense interest in treatment: the years 1880–1910, already examined in the present study, and the revival of premature infant care following the incorporation of oxygen into the incubator in the mid-1930s. Hess acted as a kind of bridge between the two periods. He functioned mainly as a consolidator, preserving and integrating the techniques developed during the first incubator phase.

For later neonatologists, the Hess nursery came to embody a conservative approach to premature infant care dominated by nurses rather than physicians. At an early stage, Hess emphasized the necessity of employing experienced nurses, rather than rotating pediatric nursing students.".[111] His essential ally in this regard was Evelyn Lundeen, who arrived in 1924 and became nursing director of the station.[112] She proved to be a tireless worker who oversaw every aspect of the nursery, developing standards of care and detailed protocols addressing virtually any situation that might arise. Under her direction, nurses assumed practical control of the nursery, often provoking resentment from younger physicians who sought to work with Hess. One doctor recalled her as "an autocrat who knew more about the care of the premature than the doctors did, and woe unto them who dared to write orders."[113] In many ways, Lundeen's nursery functioned as a no-man's-land in a literal sense. Such indictments suggest that she was indeed a powerful advocate of her profession who defied any suggestion of subservience. There may also have been friction between resident doctors and the philosophy of moderation for which Lundeen became renowned. Throughout her life, she insisted on handling premature infants with great care, and no more than necessary. Her basic convictions remained intact through the rising excitement over oxygen therapy in the 1940s, by which time she would become coauthor of Hess's second textbook, *The Premature Infant: Its Medical and Nursing Care.*[114]

Although the Hess-Lundeen nursery primarily acted as a bridge between two eras in terms of preserving the older techniques originating at turn of the century, it also anticipated an important feature of the new style of premature infant care that would appear in the 1940s. For it was Julius Hess who first transformed the incubator into an oxygen chamber. He did so, in 1934, by creating a metal hood with a small window designed to be fitted over the incubator bed. Though an extended consideration of this device would go beyond our present concerns, it did represent a significant innovation in maintaining a constant elevated pressure of oxygen around the infant. The Hess Oxygen Bed was in many ways a crude device that made observation of the baby difficult and could not maintain a high oxygen concentration. Yet it did produce superior results and helped ignite a wave of enthusiasm for oxygen therapy that would lead to much more elegant incubators capable of delivering oxygen concentrations close to 100 percent.[115] The incubator was reborn in the process, this time as an oxygen chamber. The ideal of the artificial environment was resurrected once again.

§

The years 1910–22 witnessed a retreat of the American medical profession from the care of the premature infant. As had been the case in France, the awakening of public health consciousness redirected attention from treatment to prevention. It was among infant mortality reformers, moreover, that a maternal approach akin to that of France became dominant in the United States by the time of the First World War. Differences, it should be noted, remained between the two nations. The pendulum swung more dramatically away from treatment in the United States, retreating all the way to traditional domestic means of therapy under the direction of the mother. And the American maternal reform strategy focused more on education than economic support. Nonetheless, on the subject of the premature infant the two countries were in many ways converging toward a consensus emphasizing prevention and maternal education. The distinctive aspects of the French approach broke down still further under the stress of war. In its aftermath, France no longer played a preceptorial role to the United States. The next revival of premature infant care would take place with much less input from abroad.

Although overshadowed by the efforts of obstetricians and infant mortality workers on behalf of prenatal care, pediatricians slowly made headway in staking their own claims to the premature infant during this same period. Their efforts would finally bear fruit in the late 1930s, chiefly through the leadership of Julius Hess and Evelyn Lundeen in Chicago. But the later story of the incubator's return could not have been predicted when Hess opened his station in 1922. In his own time he was an exception to the general rule of professional apathy regarding the treatment of premature infants. His research, meticulously documenting the outcome of his patients, was fundamentally defensive. Nursing procedure, representing the codification of techniques developed earlier, was basically conservative. The opening of the Hess incubator station in 1922 represented a shift from a period characterized by technology transfer and innovation to one marked by standardization and consolidation.

CONCLUSION

A Technology That Misfired?

The early history of the incubator poses problems for those who would see medical technology as evolving along a line of progress. One conclusion emerging from its analysis, however obvious, is that technology is not an autonomous agent of social change. The creation of the incubator, the first proclaimed treatment for the premature infant, resulted in a short-lived therapeutic fad rather than the birth of neonatal medicine. Two questions may remain in the reader's mind after considering its convoluted story. Why did the device fail to win acceptance? And what can its failure teach us?

The temptation implicit in the question of why the incubator failed is to see its invention as a great idea that went unappreciated by contemporaries. The preceding account does not support such a view. Far from encountering apathy or ignorance, the incubator attracted some of the most talented pediatricians and obstetricians of its time. Medical professionals with widely respected credentials took opposing sides on its value. Moreover, it is quite possible that both may have been right, depending on the setting in which they worked. Though Tarnier seemed to show that the introduction of warm incubators on the cold wards of the Maternité reduced infant mortality, American pediatricians working in infant hospitals later convinced themselves that the *rejection* of the device led to the same result. Indeed, it is very hard to sort out their opposing claims without appealing to present-day theory and practice. In this sense, the question of the incubator's rejection implies a presentist perspective. The rejection of the incubator is a "problem" because physicians today believe that incubators work. If incubators had never regained professional favor, the relevant problem would be to explain why they lasted as long as they did. Indeed, to state the question in this way is in itself misguided.

The more appropriate question for the historian is to ask how the incubator's story can illuminate the evolution of medical technology—in particular, the role played by social context in shaping its development. This is not to say that social factors determined its evolution to the extent that scientific and technical issues may be ignored. American physicians regularly denounced the technical shortcomings of the incubator. Their attacks centered to some extent on its thermostat and heating system but especially on its lack of provision for adequate ventilation. The great problem in evaluating these assertions is that they were not shared by French obstetricians. Pierre Budin, for example, turned the labor-intensive heating system of the Tarnier incubator into a virtue, claiming that it recruited the mother into an active role in her child's care. Even more striking was the fact that the French obstetricians virtually never complained about the problem of ventilation. How could the perceptions of American and French physicians have been so different? It is here that the idea of the interpretive flexibility of artifacts provides great insight. In a real sense, physicians in different settings did not see the same thing when they gazed at the incubator—or, for that matter, inside it.

More specifically, social context shaped physicians' understanding of the premature infant, which in turn influenced how they developed its treatment. This study has emphasized how three particular layers of social context (cultural, professional, and institutional) nested on top of one another to produce different perspectives on premature babies. Their effects are most clearly illustrated by the poles defined by French obstetricians and American pediatricians, the dominant forces within neonatal medicine in the two countries. We have argued that the two approached the premature newborn from opposing directions: French obstetricians moved from the mother to the newborn, American pediatricians from older to younger infants. The former more readily grasped the distinct aspects of the premature infant, in particular its peculiar vulnerability to cold exposure during the first hours and days of life. American pediatricians, based in infant rather than maternity hospitals, typically saw a relatively older population of premature infants, brought in from the home when its resources had failed and prone to infection and malnutrition as much as cold exposure. In short, premature babies appeared analogous to other weak and malnourished infants.

These two perspectives on the premature infant resulted in two different notions of the incubator. French obstetricians envisioned the device as a warming apparatus patterned after egg incubators; American pediatricians, as a comprehensive artificial environment that addressed all of the necessary ingredients thought to sustain the vitality of a living organism. In the hands of the latter profession, heating therefore lost priority to ventilation and protection from stimulation as the main functions of the apparatus. The incubator became a brooder.

Of course, it would be far too simple to see the early history of the incubator in terms of this fundamental dichotomy between the obstetricians of French maternity hospitals and the pediatricians of American infant hospitals. What adds much of the color and complexity to the story is the fact that many of its protagonists cannot be fully assigned to either role. Each country, for example, gave rise to traditions of newborn medicine outside the dominant professional culture. These were frequently harder to characterize, being influenced by their own context as well as that of the dominant specialty. French pediatricians valued fresh air in premature infant therapy, as did their American counterparts, but did not go so far as to discard the closed incubator. The American obstetrician Joseph DeLee shared the French emphasis on providing warmth to the infant early in life as well as on breast-feeding but placed as much importance on ventilation as did any pediatrician. Still finer distinctions could be drawn as well. Pierre Budin the obstetrician supervised for a short time the *service des débiles*, temporarily affording him a perspective more akin to pediatricians and leading him to concentrate on larger premature babies. The categories of culture, profession, and institution provide a framework for understanding these various responses; they did not determine them. Indeed, some of the most creative responses to premature infant care took place in the disputed territories between those of the dominant professional cultures.

Nonetheless, it is essential to emphasize the role played by social factors in the incubator's interpretation in order to understand its evolution. The problem remains of integrating them into an explanation of how medical professionals eventually did reach consensus on the device. Social constructionists have been criticized for reducing the history of technology to a kind of power struggle between opposing groups. If rival social groups interpret a given artifact in incompatible ways, the most powerful will be expected to impose its conception on the others. Such explanations have tended to overlook the role of formal evaluation procedures in the development of technology. Medical technology, however, may well represent a particularly fruitful area for their application, given the difficulties inherent in determining whether a new therapy "works" on human beings.

In the example of the incubator, French and American physicians did not develop evaluation procedures capable of overcoming the perspective imposed by their social contexts. One might object that the French did rely more heavily on statistics, whereas the Americans often cited anecdotal experience. For both groups, nonetheless, the definition and nature of the target population for the incubator to a great extent determined its results. To the degree that the impact of the incubator in a maternity hospital was fundamentally different from that in an infant hospital, American pediatricians had reason to question the relevance of statistical studies based in the former. The point was not that either the French or Americans possessed a more objective method by which to test their technology but that they simply had different evaluation procedures that probably modified the rate of technological change more than its outcome. American attitudes toward the incubator, compared with those in France, displayed a tendency to swing from one extreme to another. While the latter plodded along with the simple Tarnier *couveuse,* the former vacillated from an eager embrace of the complex Lion incubator to the outright rejection of all closed incubators within a few years. To the extent that large statistical series require more time than anecdotal experience, their use underwrites a conservative style of technological change, buffering it against erratic shifts in fashion. But statistics did not set the French obstetricians free of their social context.

If formal evaluation played a secondary role in determining professional responses to the incubator, can the evolution of the incubator be reduced to a power struggle between rival social groups? In this study we have repeatedly characterized individual physicians in terms of their professional and social context and have traced the outlines of two perspectives shaping their outlook. The French obstetric emphasis on early prematurity and temperature promoted simple incubators; the American pediatric ideal of the "artificial environment," modeled on the treatment of the premature as feeble infant, promoted a succession of responses, including the Rotch brooder, the incubator room, and the Hess heated bed and incubator station. In one sense, the different paths followed by the incubator in the two countries matched the respective domination of these two groups. French obstetricians made a more sustained bid to supervise the newborn than did their American counterparts, and pediatricians consequently entered the realm of neonatal medicine earlier in the United States than in France.

But the story still cannot be reduced to a conflict between pediatric and obstetric approaches to the newborn. These different orientations represented ideal types, not standards of practice pervading an entire specialty. Many of the most innovative figures in this study moved to some extent outside their own tradition. In the United States, pediatricians and obstetricians did not engage in a systematic power struggle over control of the newborn until after the incubator's demise. The incubator story involved not a battle between two armies but a series of forays led by their scouts into the unexplored territory represented by the newborn. And to push the frontier metaphor further, it would be wrong to portray that territory as unoccupied.

A better way to envision the social relations shaping the incubator's evolution is to place the pediatrician and obstetrician at two corners of a triangle, advancing toward a third point occupied and actively maintained by the mother. Gender thus emerges as a critical variable in our analysis, one that seems to have operated to some extent independently of the other cultural, professional, and institutional factors. Both obstetric and pediatric specialists appealed to science to justify the extension of their expertise into a domain traditionally dominated by women and the practices of domestic hygiene. The interaction involved varying degrees of cooperation and conflict, depending on the setting. Paris maternity hospitals afforded obstetricians a context where they had great power to supervise the mothers in their care. Conflict between physicians and mothers rarely erupted until the creation of *services des débiles* that admitted infants from outside. Obstetricians such as Budin saw the mother as an antagonist to their efforts in this setting and withdrew to the maternity hospital environment. In that setting, they articulated an approach that remained more domestic than scientific, relying on breast milk and simple, wood-and-glass incubators. The bedside incubator became an extension of the mother, designed to encourage her active involvement. It embodied a maternal approach to the newborn that envisioned the physician's task as one of supervising the mother in caring for her infant.

American pediatricians, in contrast, set out on a more ambitious project, to care for the newborn directly with the aid of scientific medicine. They aimed to improve upon as well as to imitate nature. Just as pediatricians developed artificial formula as a substitute for breast milk, they conceived of the incubator as an artificial environment. To a certain extent, this conception made them sympathetic with the more complex versions of the incubator that took on the functions of artificial womb, home, and nurse. In a few cases, pediatricians developed these metaphors to create new therapeutic approaches, such as the incubator room and the Hess incubator station. But the dominant theme of the pediatric profession after the early 1900s was one of retreat from the premature infant, with respect first to artificial feeding and then to the incubator. By the First World War, they had reached a position not far from that of the French, delegating premature infant care to the mother and the tradition of domestic hygiene.

The fact that the medical professions in both countries retreated from their most ambitious goals illustrates another important point about the role of gender in this story. The efforts of physicians to establish a role for themselves in supervising the newborn cannot be caricatured in terms of a self-righteous brand of medical imperialism. In contrast to the exuberance of the most outspoken incubator advocates was a remarkable degree of ambivalence among their colleagues. Here, the symbolic qualities of the incubator could be turned against it, linking it to other projects in which physicians could be criticized for overstepping their bounds in the name of science. The incubator became an artificial environment of a sinister sort, akin to the foundling hospital or urban tenement. Its fate illustrates how technology can become a symbol as well as a scapegoat for the social and moral conflicts accompanying scientific medicine.

The tension between the machine and the mother was eventually reconciled by introducing the professional nurse. This development for the most part awaited the years after 1922 but was already anticipated in the incubator stations associated with DeLee and Hess. Such physicians rejected the pretensions of the incubator itself to be a "mechanical nurse" but affirmed that it required the attention and expertise of a genuine professional nurse. During much of the second quarter of the century, as the Hess incubator station rose in prominence and began to be imitated, nurses consolidated control of the premature infant nursery. Their rise to power within this institution was notable when set against mothers as well as physicians. It raises the possibility, yet to be explored, that the professional nurse made possible what the incubator had not: the usurpation of maternal responsibility for the critically ill newborn.

A final point concerns the implications of the incubator story for newborn intensive care today. The main thrust of this study has been to explore how social context fundamentally shapes the ways physicians understand and treat their patients. A contrary theme has been the tendency of many of the physicians examined here to deny any such constraints. One thinks particularly of those American enthusiasts dazzled by the promise of scientific medicine at the turn of the century: Thomas Rotch designing an incubator that embodied the best principles of science, regardless of cost; John Zahorsky conducting scientific studies in the environment of a world's fair, and Joseph DeLee donating his own money to create his elaborate incubator station. The zeal and dedication of these individuals enabled them to create new and creative approaches to premature infant care.

Yet underlying the idealism of early twentieth-century scientific medicine lay a disquieting tendency to deny any limits to the American incubator campaign. The artificial environment became a moral, as well as mechanical, metaphor; few of the incubator's advocates in the United States shared the French concern over the need to support the infant (and mother) beyond the acute period. American neonatal medicine thus began to excel in the hospital care of premature infants, in contrast with the emphasis on prenatal and posthospital care found in France.

The social contexts that shaped the incubator are obviously not identical to those accompanying the evolution of modern newborn intensive care technology. Even so, it is essential to understand the role of longer-term historical forces in the evolution of the latter. The shift of premature infant care from the home to the hospital, as well as from the mother to the physician and nurse, took place early in the century. Of still greater importance, it was during these same years that American physicians began to define neonatology principally as a scientific challenge rather than a public health crusade. Technology thus assumed center stage in American newborn medicine at a very early point. One can argue that it continues to do so today. Most of the public debate over newborn intensive care has focused on the challenges and dilemmas posed by technology within the hospital context, overshadowing the problems of preventing premature birth or supporting infants after leaving the hospital. As a new era of economic constraint approaches, it remains to be seen whether some form of the integrated vision of newborn care associated with Pierre Budin can be adapted to the context of American medicine at the end of the twentieth century.

NOTES

Introduction

1. Technology became a particularly controversial area of medicine in the 1970s; for an overview, see the articles in "Doing Better and Feeling Worse: Health in the United States," *Daedalus* 106 (1977). Concerning the role of new technologies in shaping the bioethics movement, see David J. Rothman, *Strangers at the Bedside: A History of How Law and Bioethics Transformed Medical Decision Making* (New York: Basic Books, 1991), especially 148–89.

2. Lynn White, *Medieval Technology and Social Change* (Oxford: Clarendon Press, 1962). Regarding more recent scholarship emphasizing technology as a primary agent of social change, see Langdon Winner, *Autonomous Technology: Technics-Out-of-Control as a Theme in Political Thought* (Cambridge: MIT Press, 1977).

3. One of the most influential early studies in this regard was Elting E. Morison's essay "Gunfire at Sea: A Case Study of Innovation," in his *Men, Machines, and Modern Times* (Cambridge: MIT Press, 1966). For a survey of recent historiography on the relationship between technology and society, see the introductory essay in *The Social Shaping of Technology: How the Refrigerator Got its Hum,* ed. Donald MacKenzie and Judy Wajcman (Philadelphia: Open University Press, 1985), 2–25.

4. For an introduction, see *The Social Construction of Technological Systems: New Directions in the Sociology and History of Technology,* ed. Wiebe E. Bijker, Thomas P. Hughes, and Trevor J. Pinch (Cambridge: MIT Press, 1987).

5. Trevor J. Pinch and Wiebe E. Bijker, "The Social Construction of Facts and Artifacts; Or How the Sociology of Science and the Sociology of Technology Might Benefit Each Other," in Bijker, Hughes, and Pinch, 17–50.

6. George Basalla, *The Evolution of Technology* (Cambridge: Cambridge University Press, 1988).

7. Alex Roland, "Theories and Models of Technological Change: Semantics and Substance," *Science, Technology, and Human Values* 17 (1992): 79–100.

8. For an excellent case study illustrating this point see Edward W. Constant, "Scientific Theory and Technological Testability: Science, Dynamometers, and Water Turbines in the Nineteenth Century," *Technology and Culture* 24 (1983): 183–98.

9. Stanley Joel Reiser, *Medicine and the Reign of Technology* (Cambridge: Cambridge University Press, 1978).

10. For an overview of historical issues related to medical technology deserving further attention, see Jonathan M. Liebnau, "Medicine and Technology," *Perspectives in Biology and Medicine* 27 (1983): 76–92. Most of the limited scholarship on the subject has dealt with diagnostic instruments; noteworthy examples include Audrey B. Davis, *Medicine and Its Technology: An Introduction to the History of Medicine* (Westport, Conn.: Greenwood Press, 1982); Christopher Lawrence, "Incommunicable Knowledge: Science, Technology, and the Clinical Art in Britain, 1850–1914," *Journal of Contemporary History* 20 (1985): 503–20; and Baron H. Lerner, "The Perils of 'X-

Ray Vision': How Radiographic Images Have Historically Influenced Perception," *Perspectives in Biology and Medicine* 35 (1992): 383–97. In comparison, therapeutic devices have received little attention, with the partial exception of electrotherapeutics. See Lisa Rosner, "The Professional Context of Electrotherapeutics," *Bulletin of the History of Medicine* 43 (1988): 64–82.

11. Joel D. Howell, "Early Perceptions of the Electrocardiograph: From Arrhythmia to Infarction," *Bulletin of the History of Medicine* 58 (1984): 83–98; Joel D. Howell, "Machines and Medicine: Technology Transforms the American General Hospital," in *The American General Hospital: Communities and Social Contexts,* ed. Diana Elizabeth Long and Janet Golden (Ithaca: Cornell University Press, 1989), 109–134; Joel D. Howell, "Diagnostic Technologies: X-Rays, Electrocardiograms, and CAT Scans," *Southern California Law Review* 65 (1991): 529–64.

12. The literature addressing the ethical and policy implications of neonatal intensive care has become voluminous since the 1970s. For two succinct introductions, see U.S. Congress, *Neonatal Intensive Care for Low Birthweight Infants: Costs and Effectiveness,* OTA-HCS-38 (Washington, D.C.: Office of Technology Assessment, 1987); "Imperiled Newborns: A Report," *Hastings Center Report* 17 (Dec. 1987): 5–32.

13. Judith Walzer Leavitt, *Brought to Bed: Childbearing in America, 1750–1950* (New York: Oxford University Press, 1986).

1. Between Fetus and Weakling

1. David J. Rothman, *Strangers at the Bedside: A History of How Law and Bioethics Transformed Medical Decision Making* (New York: Basic Books, 1991), 190–221.

2. On the Baby Doe controversy, see "Imperiled Newborns: A Report," *Hastings Center Report* 17 (Dec. 1987): 7–9; Jeanne Harley Guillemin and Lynda Lytle Holmstrom, *Mixed Blessings: Intensive Care for Newborns* (New York: Oxford University Press, 1986), 11–13. For an overview of the issues surrounding the treatment of handicapped newborns, see Robert F. Weir, *Selective Nontreatment of Handicapped Newborns: Moral Dilemmas in Neonatal Medicine* (New York: Oxford University Press, 1984).

3. Gordon B. Avery, "Ethical Dilemmas in the Treatment of the Extremely Low Birth Weight Infant," *Clinics in Perinatology* 14 (1987): 362.

4. Fabian G. Eyal, "The Small-for-Gestational-Age Preterm Infant," in *Textbook of Prematurity: Antecedents, Treatment, and Outcome,* ed. Frank R. Witter and Louis G. Keith (Boston: Little, Brown, 1993), 360–61.

5. J. C. Sinclair, "Evaluation of Neonatal Intensive Care Programs," *New England Journal of Medicine* 305 (1981): 489. In 1988, infants born before thirty-seven weeks' gestation accounted for 9.7 percent of all live births; Witter and Keith, 3.

6. Steven Baumgart, "Temperature Regulation of the Premature Infant," in *Schaffer and Avery's Diseases of the Newborn,* 6th ed., ed. H. William Taeusch, Roberta A. Ballard, and Mary Ellen Avery (Philadelphia: W. B. Saunders, 1991), 255–59.

7. The preceding summary of the problems of prematurity is loosely compiled from John Sinclair and David Tudehope, "Birth Weight, Gestational Age, and Neonatal Risk," in *Behrman's Neonatal-Perinatal Medicine: Diseases of the Fetus and Infant,* 3d ed., ed. Avroy Fanaroff and Richard J. Martin (St. Louis: C. V. Mosby, 1983),

196–205; Witter and Keith, 183–84, 279–360. Epidemiological information is provided in Bea J. van den Berg and Frank W. Oechsli, "Prematurity," in *Perinatal Epidemiology*, ed. Michael B. Bracken (New York: Oxford University Press, 1984), 71.

8. Richard A. Meckel, *Save the Babies: American Public Health Reform and the Prevention of Infant Mortality, 1850–1929* (Baltimore: Johns Hopkins University Press, 1990), 1.

9. See J. Lewis Smith's discussion of early infant mortality in his *A Treatise on the Diseases of Infancy and Childhood* (Philadelphia: Henry C. Lea, 1869), 23–24; also Meckel, 32–33.

10. [William Farr], *Vital Statistics: A Memorial Volume of Selections from the Reports and Writings of William Farr, M.D.*, ed. Noel A. Humphreys (London: Offices of the Sanitary Institute, 1885; Metuchlen, N.J.: Scarecrow Press, 1975), 189.

11. *The London Practice of Midwifery*, 4th ed. (New York: Samuel Wood and Sons, 1820), 114. On the social significance of gestational age, see Ann Ellis Hanson, "The Eight Months' Child and the Etiquette of Birth: *Obstit Omen!*" *Bulletin of the History of Medicine* 61 (1987): 589–602.

12. Judith Walzer Leavitt, *Brought to Bed: Childbearing in America, 1750–1950* (New York: Oxford University Press, 1986), 196–203.

13. [William Smellie], *Smellie's Treatise on the Theory and Practice of Midwifery*, ed. Alfred H. McClintock (London: New Sydenham Society, 1876), 2: 275.

14. Benjamin Pugh, *A Treatise of Midwifery* (London: J. Buckland, 1765), 50.

15. Ibid., 51, 133, plate 2; *A Treatise on the Diseases of Infants and Children* (London: W. Johnston, 1772), 17; Alexander Hamilton, *A Treatise on the Management of Female Complaints, and of Children in Early Infancy* (New York: Samuel Campbell, 1792), 291.

16. Alexander Hamilton, 288–92; Michael Underwood, *A Treatise on the Diseases of Childhood*, 3d ed. (London: J. Mathews, 1795), 2: 66–71; Walker Keighley, *A New System of Family Medicine, for the Use of Midwives, Mothers, and Nurses* (London: B. Crosby, 1806), 329–30; James Hamilton, *Hints for the Treatment of the Principal Diseases of Infancy and Childhood* (Edinburgh: Peter Hill, 1809), 8–12; *London Practice of Midwifery*, 75–77.

17. Alexander Hamilton, 288; also see James Hamilton, 8.

18. For accounts of the history of the Humane Societies and their techniques in the eighteenth century, see Christian Augustus Struve, *A Practical Essay on the Art of Recovering Suspended Animation* (Albany: Whiting, Backus, and Whiting, 1803), 5–13, 190–93; John Lathrop, *A Discourse before the Humane Society in Boston* (Boston: E. Russell, 1787), 18–23.

19. [John Gunn], *Gunn's Domestic Medicine, or The Poor Man's Friend* (1830; facsimile ed., Knoxville: University of Tennessee Press, 1986), 339–40. Other treatises that assumed midwives to be competent in providing mouth-to-mouth ventilation to the newborn included Samuel Bard, *A Compendium of the Theory and Practice of Midwifery* (New York: Collins and Perkins, 1807), 207; John Burns, *Popular Directions for the Treatment of the Diseases of Women and Children* (New York: Thomas A. Ronalds, 1811), 123.

20. *Treatise on the Diseases of Infants and Children* 17.

21. John Locke, *Some Thoughts Concerning Education,* 5th ed. (1705), in *The Educational Writings of John Locke,* ed. James L. Axell (Cambridge: Cambridge University Press: 1968), 116–20; Jean Jacques Rousseau, *Emile,* trans. Barbara Foxley (London: J. M. Dent and Sons, 1957), 26–27. The roots of the hardening concept go back to the ancient world; according to Soranus, the Germans routinely bathed their infants in cold water. See [Soranus], *Soranus' Gynecology,* trans. Owsei Temkin (Baltimore: Johns Hopkins University Press, 1956), 82–83.

22. William Cadogan, *An Essay upon Nursing and the Management of Children,* 9th ed. (London: Robert Horsfield, 1769), 9.

23. William Buchan, *Domestic Medicine, or The Family Physician,* 2d American ed. (Philadelphia: Joseph Crukshank, 1774), 7.

24. Examples include Pliny, quoted in Jaques Guillimeau, *Child-birth, or The Happy Deliverie of Women* (London: A. Hatfield, 1612), 2; Bard, 208; Robert Ellis, *Disease in Childhood: Its Common Causes and Directions for Its Practical Management* (London: G. Cox, 1852), 36–37.

25. Although many of the ideas involved in this orientation were quite ancient, drawing ultimately on the Aristotelian and Galenic concept of innate heat sustaining the body, their use with regard to the newborn emerged as physicians began to notice that not all survivors of resuscitation did well. For a discussion of the concept of innate heat and its relation to eighteenth-century respiratory physiology, see Everett Mendelsohn, *Heat and Life: The Development of the Theory of Animal Heat* (Cambridge: Harvard University Press, 1964), 8–26.

26. John Harley Warner, "'The Nature-Trusting Heresy': American Physicians and the Concept of the Healing Power of Nature in the 1850s and 1860s," *Perspectives in American History* 11 (1977–78): 291–324; John Harley Warner, *The Therapeutic Perspective: Medical Practice, Knowledge, and Identity in America, 1820–1885* (Cambridge: Harvard University Press, 1986), 11–36, 91–97.

27. Marshall Hall, *Asphyxia: Its Nature and Its Remedy* (London: Joseph Mallett, 1856), 27–28; Hugh L. Hodge, *The Principles and Practice of Obstetrics* (Philadelphia: Henry C. Lea, 1866), 215; Alfred Vogel, *A Practical Treatise on the Diseases of Children,* 4th German ed., trans. J. Raphael (New York: D. Appleton, 1870), 53.

28. Susan M. Reverby, *Ordered to Care: The Dilemma of American Nursing, 1850–1945* (Cambridge: Cambridge University Press, 1987), 13.

29. Sylvia D. Hoffert, *Private Matters: American Attitudes toward Childbearing and Infant Nurture in the Urban North, 1800–1860* (Urbana: University of Illinois Press, 1989), 142–47; Sally G. McMillen, *Motherhood in the Old South: Pregnancy, Childbirth, and Infant Rearing* (Baton Rouge: Louisiana State University Press, 1990), 135–37.

30. On the interpretation of domestic medical guides, see Charles E. Rosenberg, "Medical Text and Social Context: Explaining William Buchan's *Domestic Medicine,*" *Bulletin of the History of Medicine* 57 (1983): 22–42, and Rosenberg's introduction to *Gunn's Domestic Medicine,* v-xxi.

31. William Potts Dewees, *A Treatise on the Physical and Medical Treatment of Children* (Philadelphia: H. C. Carey and I. Lea, 1825), 265–66.

32. Charles E. Rosenberg, "The Therapeutic Revolution: Medicine, Meaning, and Social Change in Nineteenth-Century America," in *The Therapeutic Revolution: Es-*

says in the Social History of American Medicine, ed. Morris J. Vogel and Charles E. Rosenberg (Philadelphia: University of Pennsylvania Press, 1979), 3–25.

33. An extended contemporary discussion of hereditarian ideas by an influential infant-hygiene writer is provided by Andrew Combe, *The Management of Infancy, Physiological and Moral, Intended Chiefly for the Use of Parents*, 9th ed., rev. Sir James Clark (Edinburgh: MacLachlan and Stewart, 1860), 1–9.

34. *The Maternal Physician; A Treatise on the Nurture and Management of Infants* (Philadelphia: Lewis Adams, 1818), 21–23.

35. *The American Lady's Medical Pocket-Book and Nursery-Adviser* (Philadelphia: James Kay, Jr., and Brother, 1833), 157.

36. David Holmes, *The Child's Physician: A Popular Treatise on the Management and Diseases of Infancy and Childhood* (Providence, 1857), 16–17.

37. For a Thomsonian example, see Daniel H. Whitney, *The Family Physician and Guide to Health* (Pennsylvania: H. Gilbert, 1833), 98.

38. Joel Shew, *Children: Their Hydropathic Management in Health and Disease* (New York: Fowlers and Wells, 1852), 82.

39. Vladislav Kruta, "William Frédéric Edwards," in *Dictionary of Scientific Biography*, ed. Charles Coulston Gillispie (New York: Charles Scribner's Sons, 1971), 4: 285–86.

40. W[illiam]. F. Edwards, *On the Influence of Physical Agents on Life*, 2d ed., trans. [Thomas] Hodgkin and Dr. Fisher (Philadelphia: Haswell, Barrington, and Haswell, 1838), 75–76.

41. Edwards, 73–75.

42. Cited in appendix to Edwards, 216–17.

43. Combe, *Management of Infancy*, 83. The popularity of Combe's guide was attested by its having gone through three editions and 4,250 copies in the three years after its first appearance in the early 1840s; the author reported it to have been received equally favorably in the United States at the same time. Andrew Combe, *A Treatise on the Physiological and Moral Management of Infancy, for the Use of Parents*, 4th ed. (Edinburgh: MacLachan, Stewart, 1844), vii.

44. For other examples, see Combe, *Management of Infancy*, 60; Richard T. Evanson and Henry Maunsell, *A Practical Treatise on the Management and Diseases of Children* (Philadelphia: Haswell, Barrington, and Haswell, 1838), 26; Holmes, iii; and Ellis, 35.

45. For a broader discussion of religious motives behind the sanitarian movement, see Charles E. Rosenberg and Caroll Smith-Rosenberg, "Pietism and the Origins of the American Public Health Movement: A Note on John H. Griscom and Robert M. Hartley," *Journal of the History of Medicine and Allied Sciences* 23 (1968): 16–35.

46. For accounts of the treatment of premature infants before the incubator, see John Bartlett, "The Warming Crib," *Chicago Medical Journal and Examiner* 54 (1887): 449; R. B. Gilbert, "The Care of Premature Infants after Induced and Accidental Labors," *Transactions of the Kentucky Medical Society*, n.s., 5 (1896): 189–90; William H. Taylor, "Some Points in Relation to Premature Children," *American Journal of Obstetrics and Diseases of Women and Children* 20 (1887): 1025; T. Gaillard Thomas, "The Induction of Premature Delivery as a Prophylactic Resource in Midwifery," *New York*

Medical Journal 10 (1870): 468; and Pierre Budin, *The Nursling: The Feeding and Hygiene of Premature and Full-Term Infants,* trans. William J. Maloney (London: Caxton Publishing, 1907), 9.

47. For an overview, see *Hyaline Membrane Disease: Pathogenesis and Pathophysiology,* ed. Leo Stern (Toronto: Grune and Stratton, 1984).

48. George D. Sussman, *Selling Mother's Milk: The Wet-Nursing Business in France, 1715–1914* (Urbana: University of Illinois Press, 1982), 22. Rachel Fuchs has argued that several social factors contributed to the high rate of infant abandonment in France: an increase in illegitimate births accompanying the increasing poverty of the late 1700s, the rising traffic of abandoned babies sent from the country to the city, and the ease with which infants could be abandoned; see her *Abandoned Children: Foundlings and Child Welfare in Nineteenth-Century France* (Albany: State University of New York Press, 1984), 10.

49. Fuchs, 147.

50. Charles Michel Billard, *Traité des Maladies des Enfans,* 2d ed. (Paris: J. B. Bailliere, 1833), 73–77.

51. Edward Jörg, *Die Fötuslunge im gebornen Kinde* (1835), discussed in Charles West, *Lectures on the Diseases of Infancy and Childhood* (Philadelphia: Lea and Blanchard, 1850), 149.

52. Thomas H. Tanner, *A Practical Treatise on the Diseases of Infancy and Childhood,* 2d American ed. (Philadelphia: Lindsay and Blakiston, 1866), 140; also see Johann Steiner, *Compendium of Children's Diseases: A Handbook for Practitioners and Students,* 2d ed., trans. Lawson Tait (New York: D. Appleton, 1875), 158–59; Vogel, 54–55.

53. This development mirrored a contemporary American trend toward increasing reliance on stimulants instead of depletive therapy. See Warner, *Therapeutic Perspective,* 83–161.

54. Eustace Smith, *On the Wasting Diseases of Infants and Children* (London: James Walton, 1868), 11–12.

55. West, 154.

56. Ellis, 43.

57. Vogel, 55; Tanner, 141.

58. Charles E. Rosenberg, "The Bitter Fruit: Heredity, Disease, and Social Thought," in *No Other Gods: On Science and American Social Thought* (Baltimore: Johns Hopkins University Press, 1961; 1976), 46–47.

59. The classic account of social Darwinism remains Richard Hofstadter, *Social Darwinism in American Thought* (Philadelphia: American Historical Association, 1955).

60. William Taylor, 1022.

61. S. Marx, "Incubation and Incubators," *American Medico-Surgical Bulletin* 9 (1896): 311. American physicians through the 1920s asserted that congenital syphilis was a leading cause of premature birth; see Julius H. Hess, *Premature and Congenitally Diseased Infants* (Philadelphia: Lea and Febiger, 1922), 320.

62. See John Thorne Crissey and Lawrence Charles Parish, *The Dermatology and Syphilology of the Nineteenth Century* (New York: Praeger Publishers, 1989), 91–93, 226–30.

63. On the inheritance of syphilis, see Elizabeth Lomax, "Infantile Syphilis as an

Example of Nineteenth-Century Belief in the Inheritance of Acquired Characteristics," *Journal of the History of Medicine and Allied Sciences* 34 (1979): 23–39.

64. Alfred Fournier, *Syphilis and Marriage,* trans. P. Albert Morrow (New York: D. Appleton, 1881), 55.

65. J. Parrot, "La nourricerie de l'Hospice des Enfants Assistés," *Bulletin de l'Académie de Médecine* 2d ser., 11 (1882): 839–53; Fuchs, 139. The greater willingness of Paris hospitals to care for syphilitic infants, it should be noted, occurred in the context of widespread anxiety in late-nineteenth-century France over venereal infection. See Jill Harsen, "Syphilis, Wives, and Physicians: Medical Ethics and the Family in Late-Nineteenth-Century France," *French Historical Studies* 16 (1989): 72–95.

66. "Feeding Syphilitic Infants," *Medical Record* 24 (1883): 176.

67. Henry Bettmann, "Premature Labor and the New-Born Child," *American Journal of Obstetrics and Diseases of Women and Children* 25 (1892): 322.

68. S. Waterman, "Revivification," *Medical Record* 23 (1883): 236. For another example, see J. T. Savory, "An Asphyxiated Newly-born Infant Restored by 'Marshall Hall's Ready Method' of Treatment," *Lancet* 1 (1857): 47.

69. T. T. Kirk, "A Young Living Foetus," *Medical Record* 36 (1889): 125; W. J. Hill, "Premature Labor before the Sixth Month: Survival of the Child for One Week," *Medical Record* 46 (1894): 344; C. G. Hubbard, "The Birth of a Very Small Living Child," *New York Medical Journal* 51 (1890): 491; H. C. Allen, "Early Viability of Twin Fetuses," *New York Medical Journal* 58 (1893): 207–8.

70. The British physiologist Marshall Hall probably deserves primary credit for initiating the new wave of procedures. Using intubated cadavers connected to a barometric tube, he demonstrated that by rolling back and forth the shoulders of a prone body, one could induce the intake of thirty cubic inches of air. Hall, 21–28, 34–36; Marshall Hall, "On a New Method of Effecting Artificial Respiration," *Lancet* 1 (1856): 229. Hall, it might be added, represented an anatomical tradition of physiology especially characteristic of mid-nineteenth-century Britain; see Gerald L. Geison, *Michael Foster and the Cambridge School of Physiology: The Scientific Enterprise in Late Victorian Society* (Princeton: Princeton University Press, 1978), 18–32; W. J. O'Connor, *Founders of British Physiology: A Biographical Dictionary, 1820–1885* (Manchester, U.K.: Manchester University Press, 1988), 14–17. Another important early method was that of Henry Silvester, who also used experimental apparatus to promote his own alternative maneuver, consisting of extending the arms and drawing them back down to compress the chest; see Henry R. Silvester, "A New Method of Resuscitating Still-born Children, and of Restoring Persons Apparently Drowned or Dead," *British Medical Journal* (1858): 577–79.

71. On methods for the treatment of asphyxia, see Tanner, 139–40; Vogel, 53–54; J. Whitridge Williams, *Obstetrics: A Textbook for the Use of Students and Practitioners* (New York: D. Appleton, 1903), 751–53. Examples of the use of alternating hot and cold water (and disregard for maintaining warmth) are found in W. D. Bizzell, "Prolonged Asphyxia in the New-born Infant," *American Journal of Obstetrics and the Diseases of Women and Children* 18 (1885): 180; W. E. Forest, "Resuscitation of an Infant after a Long Period of Asphyxia," *Medical Record* 16 (1879): 597–98; C. H. Humphreys, "Treatment of Asphyxia of the New-born: Amyl Nitrite," *Medical Record* 17 (1880): 553–54.

72. For a contemporary review of the relation between viability and infanticide law, see J. C. Garrigues, "Asphyxia in New-born Children Considered from a Medical and a Legal Standpoint," *American Journal of Obstetrics and the Diseases of Women and Children* 6 (1878): 800–808. On infanticide and child neglect in the mid-nineteenth century, see George K. Behlmer, "Deadly Motherhood: Infanticide and Medical Opinion in Mid-Victorian England," *Journal of the History of Medicine and Allied Sciences* 34 (1979): 403–27. Meckel, 30–31.

73. Abraham Jacobi summarized European law on viability in the discussion following Garrigues, 808; also see Alfred S. Taylor, *Medical Jurisprudence*, ed. R. Egglesfeld Griffith (Philadelphia: Lea and Blanchard, 1845), 486–87; Theodric Romeyn and John B. Beck, *Elements of Medical Jurisprudence*, 12th ed., rev. C. R. Gilman (Philadelphia: J. B. Lippincott, 1863), 1: 407–11.

74. The British General Registration Office made birth registration compulsory in 1874 and established penalties for burying live-born children as stillborn; see Alison MacFarlane and Miranda Mugford, *Birth Counts: Statistics of Pregnancy and Childbirth* (London: Her Majesty's Printing Office, 1984), 29. In the United States, accurate birth registration in the late nineteenth century was limited for the most part to Massachusetts; see James H. Cassedy, *Medicine and American Growth, 1800–1860* (Madison: University of Wisconsin Press, 1986), 175–77; U.S. Dept. of the Interior, Census Office, *Report on Vital and Social Statistics in the United States at the Eleventh Census: 1890;* pt. I, *Analysis and Rate Tables* (Washington, D.C.: Government Printing Office, 1896), 21. On the problem of live infants registered as stillbirths, see Abraham Jacobi, "Address," *Transactions of the American Association for the Study and Prevention of Infant Mortality* 1 (1910): 43.

75. W. Tyler Smith, "On the Abolition of Craniotomy from Obstetric Practice, in all Cases Where the Foetus is Living and Viable," *Transactions of the Obstetric Society of London* 1 (1859): 21–50. On craniotomy, see Jane B. Donegan, *Women and Men Midwives: Medicine, Morality, and Misogyny in Early America* (Westport, Conn.: Greenwood Press, 1978), 42–45.

76. Thomas, 451–53; W. M. Polk, "The Induction of Premature Labor," New York *Medical Journal* 48 (1888): 283–86.

77. Hodge, 289; Polk, 283.

78. A. H. Burbank, "The Induction of Premature Labor," *Medical News* 43 (1883): 18–19.

79. William Thompson Lusk, *The Science and Art of Midwifery* (New York: D. Appleton, 1883), 333.

80. Thomas, 468. More typical are the very brief discussions by Polk, 285; Lusk, 333; and Burbank, 19.

2. Tarnier's Invention

1. Thomas E. Cone, *History of the Care and Feeding of the Premature Infant* (Boston: Little, Brown, 1985), 38–39.

2. Tarnier's "ingeniosité" was especially stressed by Maurice Japiot, *Tarnier (1828 à 1897): Sa vie et son oeuvre obstétricale* (Paris: G. Steinheil, 1907), 61. Tarnier's intern at the time the incubator was introduced emphasized his mentor's diligence and perse-

verance in refining the device; see Paul Bar, "Le Professeur S. Tarnier," *La presse médicale* (1897), 2d sec., no. 102: 351.

3. Japiot, 18.

4. Pierre Budin, "Décès de M. Tarnier," *Bulletin de l'Académie de Médecine*, 3d ser., 38 (1897): 500–504; Cone, *History of the Premature Infant*, 38; Harold Speert, *Obstetric and Gynecologic Milestones* (New York: Macmillan, 1958), 481–89.

5. The earliest published accounts of the Tarnier incubator are found in H. Napias and A. J. Martin, *L'étude et les progrès de l'hygiene en France de 1878 à 1882* (Paris: G. Maisson, 1882), 308–9; Alfred Auvard, "De la couveuse pour enfants," *Archives de tocologie des maladies des femmes et des enfants nouveau-nés* 10 (1883): 577–609.

6. Auvard, "De la couveuse pour enfants," 594.

7. Cone, *History of the Premature Infant*, 24–32; Pierre Budin, *The Nursling: The Feeding and Hygiene of Premature and Full-Term Infants*, trans. William J. Maloney (London: Caxton Publishing, 1907), 10–13.

8. Carl Credé, "Ueber Erwärmungsgeräthe für frühgeborene und schwächliche kleine Kinder," *Archiv für Gynaekologie* 24 (1884): 128–47; N. Th. Miller, "Die Frühgeborenen und die Eigenthümlichkeiten ihrer Krankeiten," *Jahrbuch für Kinderheilkunde* 25 (1886): 179–94.

Thomas Cone and F. F. Marx have traced the origins of the *warmwänne* to the warming tub described in 1837 by Johann Georg von Reuhl in the Imperial Foundling Hospital in St. Petersburg. Its subsequent history is difficult to follow, since physicians, to the extent they discussed the device in writing, at all, rarely displayed any awareness of its origins. The French claimed priority for the first case report published regarding the device, by the Bordeaux physician Jean-Paul Denucé in 1857. See F. F. Marx, *Die Enwicklung der Säuglingsinkubatoren: Eine, Medizintechnische Chronik* (Bonn: Verlag Siering, 1968), 26; Thomas E. Cone, "The First Published Report of an Incubator for Use in the Care of the Premature Infant (1857)," *American Journal of Diseases of Children* 135 (1981): 658–60.

9. Credé, "Erwärmungsgeräthe," 142; Pierre Budin, "Des soins à donner aux nouveau-nés en etat de faiblesse congénitale," *La semaine médicale* 8 (1888): 193.

10. Carl Credé, "Die Verhütung der Augenentzündung der Neugeborenen," *Archiv für Gynaekologie* 17 (1881): 50–53.

11. Paul Berthod, *Les enfants nés avant terme: La couveuse et la gavage à la Maternité de Paris* (Paris: G. Rougier, 1887), 31–32, 42; Auvard, "De la couveuse pour enfants," 580–81.

12. Auvard, "De la couveuse pour enfants," 602–5; Budin, *Nursling*, 10–11.

13. Mme. Henry, "Fondation du pavillon des enfants débiles á la Maternité de Paris," *Revue des maladies de l'enfance* 16 (1898): 144; Budin, *Nursling*, 12–13.

14. Paul H. Perlstein, "Physical Environment," in *Behrman's Neonatal-Perinatal Medicine: Diseases of the Fetus and Infant*, 3d ed., ed. Avroy Fanaroff and Richard J. Martin (St. Louis: C. V. Mosby, 1983), 266; Edmund Hey and Jon W. Scopes, "Thermoregulation," in *Neonatology: Pathophysiology and Management of the Newborn*, 3d ed., ed. Gordon B. Avery (Philadelphia: J. B. Lippincott, 1987), 203.

15. Stéphane Tarnier, "Des soins à donner aux enfants nés avant terme," *Bulletin de l'Académie de Médecine*, 2d ser., 14 (1885): 951.

16. Marx, 26. The American pediatric leader Abraham Jacobi insisted on the beneficial effects of cool air on the lungs of premature infants; see Jacobi, *Therapeutics of Infancy and Childhood* (Philadelphia: J. B. Lippincott, 1896), 45.

17. Adolph Pinard, "Eloge de Tarnier," *La France médicale* 55 (1908): 472.

18. Numerous sociologists and increasing numbers of medical historians have found useful the concept that diseases are socially constructed. While not denying the reality of biological processes or the experience of illness, this interpretive perspective emphasizes the role played by social, cultural, and political factors in organizing and interpreting the phenomenon of illness into discrete diseases. A useful introduction is provided by Peter Wright and Andrew Treacher, *The Problem of Medical Knowledge: Examining the Social Construction of Medicine* (Edinburgh: Edinburgh University Press, 1982).

19. An important exception is Jack D. Ellis, *The Physician-Legislators of France: Medicine and Politics in the Early Third Republic, 1870–1914* (Cambridge: Cambridge University Press, 1990). The primary focus of this work, however, remains more political than medical.

20. The wide range of perspectives on the Paris clinical school are illustrated by Erwin H. Ackerknect, *Medicine at the Paris Hospital, 1794–1848* (Baltimore: Johns Hopkins Press, 1967); Michel Foucault, *The Birth of the Clinic: An Archaeology of Medical Perception*, trans. A. M. Sheridan Smith (New York: Pantheon Books, 1973); Toby Gelfand, *Professionalizing Modern Medicine: Paris Surgeons and Institutions in the Eighteenth Century* (Westport, Conn.: Greenwood Press, 1980); John E. Lesch, *Science and Medicine in France: The Emergence of Experimental Physiology, 1790–1855* (Cambridge: Harvard University Press, 1884); and Russell C. Maulitz, *Morbid Appearances: The Anatomy of Pathology in the Early Nineteenth Century* (Cambridge: Cambridge University Press, 1987).

21. Ellis, 44.

22. The demographic aspects of the Maternité are explored in Rachel G. Fuchs and Paul E. Knepper, "Women in the Paris Maternity Hospital: Public Policy in the Nineteenth Century," *Social Science History* 13 (1989): 187–209. On the hospital as a form of social welfare, see John H. Weiss, "Origins of the French Welfare State: Poor Relief in the Third Republic, 1871–1914," *French Historical Studies* 13 (1983): 47–78. Hospital history and standards are described by A. d'Echérac, *L'assistance publique: Ce Qu'elle fut, ce qu'elle est* (Paris: G. Steinheil, 1909), 126–30.

23. Japiot, 38.

24. Adolphe Pinard, *Discours: Stéphane Tarnier* (Paris: G. Steinheil, 1898), 5.

25. A useful summary of the moral and scientific issues surrounding the contagionist and miasmatic theories of infection is provided by Charles Rosenberg, "Florence Nightingale on Contagion: The Hospital as a Moral Universe," in *Healing and History: Essays for George Rosen*, ed. C. E. Rosenberg (New York: Science History Publications, 1979), 116–36.

26. Stéphane Tarnier, *De la fièvre puerpérale observée à l'hospice de la Maternité* (Paris: J. B. Baillière et fils, 1858).

27. Oliver Wendell Holmes, "On the Contagiousness of Puerperal Fever," *New England Quarterly Journal of Medicine and Surgery* 1 (1843): 503–30; Ignaz Semmel-

weis, *The Etiology, Concept and Prophylaxis of Childbed Fever,* ed. and trans. K. Codell Carter (Madison: University of Wisconsin Press, 1983).

28. Japiot, 12–17.

29. Stéphane Tarnier, "Description d'un nouveau moyen de provoquer l'accouchement prématuré" (1862), cited in Pinard, *Discours: Stéphane Tarnier,* 5; Stéphane Tarnier, "Avortement provoqué chez une femme atteinte d'ostéomalacie," *Mémoires de la Société de Chirurgie de Paris* 6 (1868): 377–85.

30. Stéphane Tarnier, "Annotation," in Paul Cazeaux, *A Theoretical and Practical Treatise on Midwifery,* 6th American ed., rev. and ann. S. Tarnier, trans. William R. Bullock (Philadelphia: Lindsay and Blakiston, 1875 [1868], 1022.

31. Cazeaux, 1004.

32. Japiot, 41; Stéphane Tarnier, "Présentations d'ouvrages, manuscrits, et imprimés," *Bulletin de l'Académie de Médecine,* 2d ser., 11 (1882): 793–94; Fuchs and Knepper, 199.

33. Pierre Budin, *Le Professeur Tarnier* (Paris: Felix Alcan, 1898), 5–7.

34. Tarnier, "Présentations d'ouvrages," 793; Alphonse Thévenot, "Rapport sur les nouvelles maternités," *Annals de hygiene publique et médecine legal* 8 (1882): 257.

35. Thévenot, 258; Japiot, 43.

36. Budin, *Le Professeur Tarnier,* 7–9; Japiot, 43.

37. Japiot, 41–42; Judith Walzer Leavitt, *Brought to Bed: Childbearing in America, 1750–1950* (New York: Oxford University Press, 1986), 159.

38. Tarnier's rise in status led to his appointment as professor of Obstetrics at the Faculty of Medicine in 1884, where he gave an opening lecture on his introduction of asepsis. Alfred Auvard, "Report of the Progress of Gynecology and Obstetrics in France," *American Journal of Obstetrics and Diseases of Women and Children* 17 (1884): 636. In 1891, he was elected president of the Academy of Medicine; see Budin, "Décès de M. Tarnier," 501.

39. Alfred Auvard, "Quarterly Report on Obstetrics and Gynecology in France," *American Journal of Obstetrics and Diseases of Women and Children* 16 (1883): 1059.

40. Stéphane Tarnier and Pierre Budin, *Traité de art de accouchements,* 2d ed., 4 vols. (Paris: G. Steinheil, 1886).

41. Pinard, "Eloge de Tarnier," 472.

42. The appearance of poultry incubators in the zoo at this point may have reflected a rise in their own popularity. Though such devices originated in Egypt and were described in Europe by the early seventeenth century, they became more widespread after the mid-nineteenth century. In France, five of the eight patents issued for poultry incubators between 1791 to 1876 were granted after 1870. U.S. Dept. of Interior, Patent Office, *Subject-Matter Index of Patents for Inventions Granted in France from 1791 to 1876, Inclusive* (Washington, D.C.: Government Printing Office, 1883), 395. On the early history of the chicken incubator, see Geoffrey Keynes, *The Life of William Harvey* (Oxford: Clarendon Press, 1966), 342–43; F. W. Gibbs, "Invention in Chemical Industries," in *A History of Technology,* ed. Charles Singer, E. J. Holmyard, A. R. Hall, and T. I. Williams (Oxford: Clarendon Press, 1957), 3:679–80.

43. Budin, "Des soins à donner aux nouveau-nés," 193.

44. Auvard, "De la couveuse pour enfants," 594.

45. Tarnier, "Enfants nés avant terme," 945.

46. Berthod, 49.

47. C. A. Wunderlich, *On the Temperature in Diseases,* 2d German ed., trans. W. Bathurst Woodman (London: New Sydenham Society, 1871). On Wunderlich's influence, see Stanley Joel Reiser, *Medicine and the Reign of Technology* (Cambridge: Cambridge University Press, 1978), 116–19.

48. In his bibliography in 1883, Auvard cited the following authors for their work on the temperature of the newborn: Andral (1870), F. V. Bärensprung (1851), H. Fehling (1874), James Finlayson (1868, 1869), R. Förster (1862), Carl Sommer (1880), and Georg Wurster (1869); see Auvard, "De la couveuse pour enfants," 594–95, 606–9.

49. Alexandre Gueniot, cited in Georges Bertin, *Contribution a l'étude des infections des nouveau-nés dans les couveuses* (Paris: G. Steinheil, 1899), 11.

50. Auvard, "De la couveuse pour enfants," 594–99.

51. Berthod affirmed the importance of maintaining a rectal temperature over 98.6°F more explicitly than did Auvard but still did not emphasize the point; Auvard, "De la couveuse pour enfants," 595–97; Berthod, 34, 38–40.

52. Tarnier, "Enfants nés avant terme," 944.

53. Berthod remarked specifically that oxygen might favorably affect certain diseases of prematurity, such as atelectasis and cyanosis; *Les enfants nés avant terme,* 44.

54. An analogy might be provided by the air pump in seventeenth-century England; see Steven Shapin and Simon Schaffer, *Leviathan and the Air-Pump: Hobbes, Boyle, and the Experimental Life* (Princeton: Princeton University Press, 1985).

55. Edward Henoch, *Lectures on Diseases of Children* (New York: William Wood, 1882), 11–32.

56. Auvard, "De la couveuse pour enfants," 587; examples include cases 74, 80, 85, and 91 in the table on 588–93.

57. An example is provided by the way in which Auvard and Berthod defined the infants who overlapped their two case series. Auvard in 1883 typically used being *avant terme* as the leading reason infants in his series required an incubator; four years later, Berthod relabeled many of the same patients as having *faiblesse congénitale.* Though Berthod also applied the full term to a limited number of infants allegedly born at full term, almost all of these weighed less than three thousand grams and thus were relatively small. His series of 578 incubator infants included 43 babies with congenital feebleness born at full term; only four of these weighed more than three thousand grams. See table in Auvard, 588–93; Berthod, 55–62.

58. For example, Budin in the 1890s preferred the term *débile* to *avant terme* but affirmed that such infants were, as a rule, the products of premature labor. Budin, *Nursling,* 2.

59. See, especially, Henoch, 356–57.

60. Budin, "Des soins à donner aux nouveau-nés," 193.

61. Pierre Budin and E. Crouzat, *La pratique des accouchements à l'usage des sages-femmes* (Paris: Octave Doin, 1891), 525.

62. Henoch, 25. Since the condition has essentially disappeared from modern neonatology, its explanation in presumably metabolic terms remains elusive. Lawrence M. Solomon and Nancy B. Esterly, "Benign Skin Disorders," in *Schaffer and Avery's*

Diseases of the Newborn, 6th. ed., ed. H. William Taeusch, Roberta A. Ballard, and Mary Ellen Avery (Philadelphia: W. B. Saunders, 1991), 994.

63. Berthod, 53.

64. Budin, *Nursling,* 5.

65. Auvard, "De la couveuse pour enfants," 585, 588–93; Berthod, 55–60.

66. On earlier controversies over the length of pregnancy, see Lindsay Wilson, *Women and Medicine in the French Enlightenment: The Debate over* Maladies des Femmes (Baltimore: Johns Hopkins University Press, 1993), 34–64.

67. Auvard, 584; Budin, "Des soins à donner aux nouveau-nés," 193.

68. John Sinclair and David Tudehope, "Birth Weight, Gestational Age, and Neonatal Risk," in Fanaroff and Martin, 197, 201.

69. Auvard cited a number of cases in the six- to seven-month range (1,400–1,820 grams) whose survival he believed clearly supported the couveuse. His emphasis on smaller infants was not simply a matter of optimistic rhetoric; this population was easier to compare because it was treated with the incubator consistently, whereas the device was used only intermittently for larger infants, according to their need; "De la couveuse pour enfants," 584, 594.

70. Berthod, 49.

71. Tarnier, "Enfants nés avant terme," 949; Tarnier and Budin, 4:523.

72. The prominent obstetrician Adolphe Pinard later became particularly critical of Tarnier's enthusiasm; Pinard, "Eloge de Tarnier," 472.

73. Stéphane Tarnier, "L'allaitement artificiel des nouveau-nés," *Bulletin de l'Académie de Médecine,* 2d ser., 11 (1882): 1090–91. Fittingly, Tarnier proposed in this article the use of comparative statistics to assess artificial feeding versus breast-feeding.

74. Quoted in Charles Maygrier, "Pierre Budin," *La France médicale* 54 (1907): 52.

3. Mothers and Nurslings

1. [Pierre Budin], "Conference de M. le Professeur Budin," 10 Nov. 1900, Dr. Paul Toubas collection, Children's Hospital of Oklahoma, Oklahoma City, 1.

2. Pierre Budin, *The Nursling: The Feeding, and Hygiene of Premature and Full-Term Infants,* trans. William J. Maloney (London: Caxton Publishing, 1907; originally *Le Nourrisson,* 1900), 57–62; "La puériculture et la pouponnière," *La nature* 25 (1897): 394; V. Hutinel and M. Delestre, "Les couveuses aux Enfants-Assistés," *Revue des maladies de l'enfance* 17 (1899): 537.

3. On the relationship between depopulation and maternal and infant health, see Rachel Fuchs, *Poor and Pregnant in Paris: Strategies for Survival in the Nineteenth Century* (New Brunswick, N.J.: Rutgers University Press, 1992), 56–76; Alisa Klaus, *Every Child a Lion: The Origins of Maternal and Infant Health Policy in the United States and France, 1890–1920* (Ithaca: Cornell University Press, 1993), 23–31. For an overview of the depopulation controversy, see Theodore Zeldin, *France, 1848–1945,* (Oxford: Clarendon Press, 1977), 2:948–69.

4. John H. Weiss, "Origins of the French Welfare State: Poor Relief in the Third Republic, 1871–1914," *French Historical Studies* 13 (1983): 56–57; Jane Ellen Crisler, "Saving the Seed; The Scientific Preservation of Children in France during the Third Republic" (Ph.D. diss., University of Wisconsin, Madison, 1984), 16.

5. Fuchs, *Poor and Pregnant,* 70; Karen Offen, "Depopulation, Nationalism, and Feminism in Fin-de-Siecle France," *American Historical Review* 89 (1984): 652–53; Crisler, 10–12.

6. Budin, "Conference," 1–4.

7. Ibid.; Fuchs and Knepper, "Women in the Paris Maternity Hospital: Public Policy in the Nineteenth Century," *Social Science History* 13 (1989): 187–88, 196.

8. H. Napias and A. J. Martin, *L'étude et les progrès de l'hygiene en France de 1878 à 1882* (Paris: G. Maisson, 1882), 309.

9. Quoted in Stéphane Tarnier, "Des soins à donner aux enfants nés avant terme," *Bulletin de l'Académie de Médecine,* 2d ser., 14 (1885): 951.

10. Mme. Henry, "Fondation du pavillon des enfants débiles à la Maternité de Paris," *Revue des maladies de l'enfance* 16 (1898): 143.

11. Ibid., 142–43; Fuchs, *Poor and Pregnant,* 142–46.

12. Stéphane Tarnier, "L'allaitement artificiel des nouveau-nés," *Bulletin de l'Académie de Médecine,* 2d ser., 11 (1882): 1090–91; J. Parrot, "La nourricerie de l'Hospice des Enfants Assistés," *Bulletin de l'Académie de Médecine,* 2d ser., 11 (1882): 852.

13. Henry, 143–44.

14. *Fécondité* was published in 1899; for an English translation, see Emile Zola, *Fruitfulness,* trans. Ernest Vizetelly (New York: Doubleday, 1900); on its impact, see Fuchs, *Poor and Pregnant,* 56–57.

15. Fuchs, *Poor and Pregnant,* 60–61.

16. Henry, 143–44.

17. Ibid.

18. Budin and Charles Maygrier in their correspondence charged that Auvard's appointment as chief resident of La Clinique, the next logical promotion following his internship at the Maternité, was blocked during his examinations by the powerful obstetrician Pajot. Pajot, incidentally, created one of the first incubator rooms as an alternative to Auvard's incubator. Auvard at any rate never obtained an obstetric appointment; he seems to have redefined himself more as a gynecologist. Charles Maygrier to Pierre Budin, 24 July 1884; Pierre Budin, *Carnet du notes,* 30–31 July 1896, Toubas collection.

19. Stéphane Tarnier and Pierre Budin, *Traité de art de accouchements,* 2d ed., 4 vols. (Paris: G. Steinheil, 1886).

20. Maurice Japiot, *Tarnier (1828 à 1897): Sa vie et son oeuvre obstétricale* (Paris: G. Steinheil, 1907), 30.

21. Mary Lynn McDougall, "Protecting Infants: The French Campaign for Maternity Leaves, 1890s-1913," *French Historical Studies* 13 (1983): 92–93.

22. Klaus, 62–64; Leonard Robinson, "Consultations for Infants in France: Their Origin, Organisation, and Results," *Practitioner* 75 (1905): 479–88; Pierre Budin, "Des consultations de nourrissons, leurs résultats," *Bulletin de l'Académie de Médecine,* 3d ser., 55 (1906): 652–58.

23. Klaus, 66–67; Robinson, 487.

24. Budin, "Allaitement et hygiène du nourrisson," *Obstétrique* 11 (1906): 1; Charles Maygrier, "Pierre Budin," *La France médicale* 54 (1907), 52.

25. For consideration of the remarkably swift conversion of hygienists and physi-

cians to the germ theory in France, see Bruno Latour, *The Pasteurization of France*, trans. Alan Sheridan and John Law (Cambridge: Harvard University Press, 1988).

26. Crisler, 40–43.

27. On positivism and anticlericalism, see Roger Magraw, *France, 1815–1914: The Bourgeois Century* (Oxford: Fontana Paperback, 1983), 192–96, 245–54. American reactions include "Expulsion of the Religious Orders from the Paris Hospitals," *Medical Record* 24 (1883): 297; and "Trouble in Hospital Management in France," *Medical Record* 25 (1884): 728.

28. Paul Strauss, "Pierre Budin," *Revue philanthropique*, 20 (1906–7): 401.

29. Maygrier, "Pierre Budin," 52.

30. Alfred Auvard, "De la couveuse pour enfants," *Archives de tocologie des maladies des femmes et des enfants nouveau-nés* 10 (1883): 594.

31. Pierre Budin, *Carnet du notes*, 24 July 1896.

32. Pierre Budin, *The Nursling: The Feeding and Hygiene of Premature and Full-Term Infants*, trans. William J. Maloney (London: Caxton Publishing, 1907), 64; Adolphe Pinard, "Rapport sur la puériculture dans ses périodes initiales," in Commission de la dépopulation, Sous-commission de la mortalité, *Séances, 1902–1903*, 5 Mar. 1902, 9 Mar. 1902 (Melun: Imprimerie Administrative, 1903), 27.

33. Budin, *Nursling*, 6–8; for extensive documentation, see Pierre Budin, "Service des enfants débiles à la Maternité, années 1896–1897," *Obstétrique* 4 (1899): 107–40.

34. Budin, "Service des enfants débiles," 113.

35. Budin, *Nursling*, 63–64.

36. [Charles] Porak and Durante, "Service des débiles à la Maternité," *Archives de médecine de l'enfance* 5 (1902): 642.

37. Pierre Budin, "Les enfants débiles," *La presse médicale*, (1902), 2d sec., no. 97: 1156.

38. Budin, *Nursling*, 6.

39. I have found little biographical information on Porak. He was listed as chief of service of the *pavillon de débiles* in 1900 in *L'assistance publique en 1900* (Paris: Generale de l'assistance publique, 1900), 500.

40. Porak and Durante, 642.

41. In the eighteenth century, the custom of sending one's infant to be nursed in the country was so prevalent that Paris and other cities were virtually devoid of infants. Though no longer as pervasive during the 1800s, wet-nursing remained common until the First World War. For a general review, see George D. Sussman, *Selling Mother's Milk: The Wet-Nursing Business in France, 1715–1914* (Urbana: University of Illinois Press, 1982).

42. Reform efforts during the first three-quarters of the century centered mainly on the moral issue of whether the hospital actually encouraged abandonment. The symbol of easy relinquishment was the *tours*, the revolving turnstile at the entrance to the hospital by means of which a parent could leave an infant with complete anonymity. The opponents of the *tours* argued that it encouraged easy abandonment; its defenders (notably the Catholic Church) maintained that it served as a protection against infanticide. The dominant critique of the foundling hospital for most of the nineteenth century thus centered on moral exhortation and the discouragement of

abandonment rather than the development of alternatives. See Rachel Fuchs, *Abandoned Children: Foundlings and Child Welfare in Nineteenth-Century France* (Albany: State University of New York Press, 1984), 21–23, 33–60.

43. On the transformation of Enfants Trouvés and the increasingly sympathetic attitude prevalent toward the end of the century see, Fuchs, *Abandoned Children,* 54–56, 149–50; for a contemporary perspective, see "La puériculture et la pouponnière," 392.

44. Porak and Durante, 644–45.

45. Fuchs, *Abandoned Children,* 126–29, 142.

46. The deWatteville Investigative Commission of 1860 concluded that most infant deaths were from "weakness" on arrival; Fuchs, *Abandoned Children,* 147.

47. Paul DeLaunay, "Traveaux et critique: Les enfants trouvés à l'hospice de la Maternité (1795–1815)," *La France médicale* 54 (1907): 337–42.

48. *Assistance publique en 1900,* 500.

49. Budin, *Nursling,* 34–50.

50. Ibid., 50.

51. Although dramatic, these epidemics accounted for only a relatively small portion of the total mortality; in 1897, the two bronchitis epidemics produced 34 out of a total of 199 deaths. Budin, "Service des enfants débiles," 126–34.

52. Budin's successor made little mention of the isolation facilities or epidemic bronchitis; Porak and Durante, 641–44.

53. Budin, *Nursling,* 68.

54. Ibid., 65–66.

55. Porak and Durante, 643–44.

56. Bernard-Jean-Antonin Marfan, *Traité de l'allaitement et de l'alimentation des enfants du premier age* (Paris: G. Steinheil, 1898), quoted in Ernst Deutsch, "Die Lage der Frühgeborenen in den Geburtsanstalten," *Archiv für Kinderheilkunde* 28 (1900): 250.

57. Georges Bertin, *Contribution a l'étude des infections des nouveau-nés dans les couveuses* (Paris: G. Steinheil, 1899), 71–73; also abstracted as "Infections of New-born Infants in the Incubator," *American Journal of Obstetrics and Diseases of Women and Children* 40 (1899): 283–85.

58. A survey taken in 1900 of thirty-four maternity hospitals (almost all European) revealed that ten used either a Tarnier or Auvard *couveuse,* seven used a double-walled warming tub, and four used a Lion incubator. The others employed no specific apparatus. Deutsch, 256–61.

59. Budin, *Nursling,* 13.

60. Budin particularly enjoyed ridiculing the opponents of asepsis; Budin, *Carnet du Notes,* 29 July 1896. The obstetric response may be compared to that of the public health community to Pasteur; see Latour, 59–110.

61. Marcel Delestre, *Etude sur les infections chez le prématuré* (Paris: G. Steinheil, 1901), 10.

62. Hutinel and Delestre, 532, 538.

63. Budin, "Allaitement et hygiène du nourrisson," 1.

64. Bertin, 73; Delestre, 139; Hutinel and Delestre, 534.

65. Parenthetically it is worth noting that the pediatrician Variot focused on the

provision of sterilized milk through milk stations (*gouttes de lait*) and did not significantly address premature infants; Klaus, 63, 82.

66. On Pinard and eugenics, see William Schneider, "Toward the Improvement of the Human Race: The History of Eugenics in France," *Journal of Modern History* 54 (1982): 268–91, especially 272–78.

67. Adolphe Pinard, "Note pour servir a l'histoire de la puériculture intrauterine," *Bulletin de l'Académie de Médecine,* 3d ser., 34 (1895): 594–95. On Pinard and maternity leave, see McDougall, 94–96.

68. Budin, "Les enfants débiles," 1156.

69. A. Belmin, "Visites de la Societé Internationale: La Clinique Tarnier et le Dr. Budin," *Revue philanthropique* 18 (1905–6): 490.

70. Budin, "Les enfants débiles," 1156.

71. Belmin, 491.

72. Budin, *Nursling,* 68.

73. By 1900, Budin mentioned allegations that premature infants were prone to Little's disease (cerebral palsy); see *Nursling,* 68–69.

74. Pinard, "Rapport sur la puériculture," 28–29.

75. Pierre Budin, "Rapport sur la mortalité infantile de 0 à 1 an," in Commission de la dépopulation, Sous-commission de la mortalité, *Séances, 1902–1903,* 12 nov. 1902 (Melun: Imprimerie Administrative, 1903), 36.

76. Budin, "Rapport sur la mortalité infantile," 38; for original report, see Charles Maygrier, "Deux statistiques hospitalières de prématurés (Lariboisière 1895–98, Charité 1898–1901)," *Obstétrique* 8 (1903): 6.

77. Pierre Budin and M. Perrett, "Nouvelle recherches sur les enfants débiles," *Obstétrique* 6 (1901): 214–15.

78. Budin, *Nursling,* 69.

79. Budin, "Rapport sur la mortalité infantile," 38.

80. Budin, *Nursling,* 14; Tarnier, "Enfants nés avant terme," 944. Part of the difference may have reflected where the thermometer was placed. Budin noted that a temperature of 77–78°F in the infant chamber corresponded to 86°F next to the hot water reservoir; contemporaries rarely described just where the temperature should be measured.

81. Budin, *Nursling,* 63–64. The French demographer J. Bertillon wrote that infants who died within the first forty-eight hours of life were often considered stillborn. Budin, "Rapport sur la mortalité infantile," 38.

82. For example, Budin's survival rate for the 1,500–2,000 gram category was roughly the same as that obtained in New York at the same time by James Voorhees at the Sloane Maternity Hospital, when deaths in the first forty-eight hours are excluded. Voorhees' hospital survival rate for this group was 71 to 82 percent, depending on whether the patients transferred to neighboring infant hospitals were counted, as compared to Budin's 76 percent survival rate. James Voorhees, "The Care of Premature Babies in Incubators," *Archives of Pediatrics* 17 (1900): 343–45.

83. Budin, *Nursling,* chaps. 1–3.

84. Pierre Budin, "La ville de Paris et la mortalité infantile," *Revue philanthropique* 14 (1903–4): 387.

85. Budin, "Les enfants débiles," 1157; Budin, "Rapport sur la mortalité infantile," 44; McDougall, 92–93.

86. McDougall, 103; Richard A. Meckel, *Save the Babies: American Public Health Reform and the Prevention of Infant Mortality, 1850–1929* (Baltimore: Johns Hopkins University Press, 1990), 100–101.

87. M. Perret, introduction to Pierre Budin, "La mortalité infantile dans les Bouches-du-Rhône," *Revue philanthropique* 20 (1906–7): 549–50.

88. "Fondation Pierre Budin," (Paris: A. Coueslant, 1909), 3, Toubas collection.

89. Charles Maygrier, "Le prématuré: sa protection, son avenir," *La presse médicale* (1902), no. 102: 819.

90. Adolf Pinard, "Eloge de Tarnier," *La France médicale* 55 (1908): 472.

91. Quoted in U.S. Dept. of Labor, Children's Bureau, "Infant-Welfare Work in Europe," by Nettie McGill (Washington, D.C.: Government Printing Office, 1921), 97, 88, 89–90.

92. Thomas E. Cone, *History of the Care and Feeding of the Premature Infant* (Boston: Little, Brown, 1985), 39.

93. See the tribute by Cone, 38–39.

4. Technology Transfer and Transformation

1. S. Marx, "Incubation and Incubators," *American Medico-Surgical Bulletin* 9 (1896): 311.

2. E. H. Grandin, "Auvard: The Incubator for Infants," *American Journal of Obstetrics and Diseases of Women and Children* 17 (1884): 421–24. Other examples include "The Couvreuse *[sic]*, or Mechanical Nurse," *Lancet* 2 (1883): 241–42, reprinted in *Medical Record* 24 (1883): 333–34; "Auvard's Couveuse, or Nest," *Transactions of the Obstetrical Society of London* 35 (1884): 25–26; "Rearing Premature Infants," *Medical Record* 28 (1885): 398; and H. C. Coe, "Credé (Leipzig): A Warming Apparatus for Premature and Weakly Children," *American Journal of Obstetrics and Diseases of Women and Children* 18 (1885): 441.

3. "Auvard's Couveuse," *Transactions of the Obstetrical Society of London* 35 (1884): 25. The Boston instrument makers Codman and Shurtleff produced the first American-manufactured version of the Tarnier-Auvard incubator; see "La Couveuse, for Preserving the Life of Feeble and Prematurely Born Infants," *Medical Record* 37 (1890): 721.

4. "The Treatment of Prematurely Born Infants," *American Journal of the Medical Sciences* 91 (1886): 329.

5. "The Couveuse pour Enfants," *Medical Record* 25 (1884): 99.

6. Paul Berthod, *Les enfants nés avant terme: La couveuse et la gavage à la Maternité de Paris* (Paris: G. Rougier, 1887), 38–40.

7. Thomas P. Hughes, "The Era of Independent Inventors," in *Science in Reflection*, ed. Edna Ullmann-Margalit (Boston: Kluwer Academic Publishers, 1988), 151–67; Thomas P. Hughes, *American Genesis: A Century of Invention and Technological Enthusiasm, 1870–1970* (New York: Viking Press, 1989), 14–15.

8. Hughes, "Era of Independent Inventors," 152.

9. Ibid., 153–54; Thomas P. Hughes, *Networks of Power: Electrification in Western Society, 1880–1930* (Baltimore: Johns Hopkins University Press, 1983), 21.

10. On the attitude of American inventors toward science, see Hughes, *American Genesis,* 47; Daniel Kevles, *The Physicists: The History of a Scientific Community in Modern America* (Cambridge: Harvard University Press, 1987), 8.

11. James R. Chadwick, "Obstetric and Gynaecological Literature, 1876–1881" *Boston Medical and Surgical Journal* 105 (1881): 247.

12. Judith Walzer Leavitt, *Brought to Bed: Childbearing in America, 1750–1950* (New York: Oxford University Press, 1986), 142–54.

13. Joel D. Howell, introduction to *Technology and American Medical Practice, 1880–1930,* ed. J. D. Howell (New York: Garland Publishing, 1988), xii.

14. Edward J. Brown, "A New Baby Incubator," *Medical Record* 41 (1892): 446–47.

15. The extensive discussion of feeding and nursing issues in Budin's *Le nourrisson* in 1900 [published as Pierre Budin, *The Nursling: The Feeding and Hygiene of Premature and Full-Term Infants,* trans. William J. Maloney (London: Caxton Publishing, 1907)] may be contrasted with A. Auvard, "De la couveuse pour enfants," *Archives de tocologie des maladies des femmes et des enfants nouveau-nés* 10 (1883): 577–609.

16. Thomas Edison, for example, developed one of his most complex inventions, the quadruplex telegraph, through the use of an analogy to a water system with pipes, valves, pumps, and waterwheels. Hughes, "Era of Independent Inventors," 158–59.

17. The longest such abstract devoted barely a sentence to Auvard's research on the incubator and infant temperature curves; Grandin, "Auvard," 424.

18. On breathing cool air as an aspect of infant hygiene, see Abraham Jacobi, *Therapeutics of Infancy and Childhood* (Philadelphia: J. B. Lippincott, 1896), 45; Vilray Papin Blair, "Premature Infants: The Necessity for and the Difficulty of Formulating a General Plan for Their Care," *St. Louis Courier of Medicine* 33 (1905): 209.

19. References to double-jacketed warming tubs prior to the 1880s were extremely sporadic; one brief reference was provided in T. Gaillard Thomas, "The Induction of Premature Delivery as a Prophylactic Resource in Midwifery," *New York Medical Journal* 10 (1870): 468. A few physicians, generally obstetricians, published brief accounts of the device after Credé's 1884 article, the most extensive of which was John Bartlett, "The Warming-Crib," *Chicago Medical Journal and Examiner* 54 (1887): 49–54. Also see Hirst, "The Incubator," *American Journal of Obstetrics and Diseases of Women and Children* 21 (1888): 631; Adelaide Brown, "Care of Premature and Feeble Infants," *Transactions of the Medical Society of California* 26 (1896): 157.

20. Franz Winckel, "Ueber Antwendung permanter Bäder bei Neugeborenen," *Contralblatt für Gynäkologie* 1 (1882): 1–38; Auvard, 599–602.

21. Auvard, 601.

22. Joseph B. DeLee, "Infant Incubation, with the Presentation of a New Incubator and a Description of the System at the Chicago Lying-in Hospital," *Chicago Medical Recorder* 22 (1902): 23–24.

23. Thomas M. Rotch, "Description of a New Incubator," *Transactions of the American Pediatric Society* 5 (1893): 44–46.

24. For Rotch's biography, see Fritz B. Talbot, "Thomas M. Rotch," in *Pediatric Profiles,* ed. Borden Smith Veeder (St. Louis: C. V. Mosby, 1957), 29–32; and Thomas E. Cone, Jr., "Thomas Morgan Rotch," in *Dictionary of American Medical Biography,*

ed. Martin Kaufman, Stuart Galishoff, and Todd Savitt (Westport, Conn.: Greenwood Press, 1984), 2: 650–51.

25. Thomas M. Rotch, *Pediatrics: The Hygienic and Medical Treatment of Children* (Philadelphia: J. B. Lippincott, 1895), 301.

26. Rotch, *Pediatrics,* 296–99, esp. 299.

27. Rotch, "Description," 44–46.

28. Vanderpoel Adriance, "Premature Infants," *American Journal of the Medical Sciences,* n.s., 121 (1901): 417. For other examples of the intrauterine metaphor, see Brown, "Care of Premature and Feeble Infants," 157–58; Lyman C. Holcombe, "The Care of Premature Infants," *Vermont Medical Monthly* 12 (1906): 51.

29. Rotch, *Pediatrics,* 309.

30. Ibid., 300.

31. "Nerves" were often cited by women as a problem with breast-feeding well into the twentieth century; see Rima D. Apple, *Mothers and Medicine: A Social History of Infant Feeding, 1890–1950* (Madison: University of Wisconsin Press, 1987), 110, 163–64.

32. George M. Beard, *American Nervousness: Its Causes and Consequences* (New York: G. P. Putnam's Sons, 1881), 96–138. See the analysis in Barbara Sicherman, "The Uses of a Diagnosis: Doctors, Patients, and Neurasthenia," *Journal of the History of Medicine and Allied Sciences* 32 (1977): 33–54.

33. Caroll Smith-Rosenberg and Charles Rosenberg, "The Female Animal: Medical and Biological Views of Woman and Her Role in Nineteenth-Century America," *Journal of American History* 60 (1973): 332–56. Regarding the extent to which neurasthenia represented a "female" affliction, see Ann Douglas Wood, "'The Fashionable Diseases': Women's Complaints and their Treatment in Nineteenth-Century America," *Journal of Interdisciplinary History* 4 (1973): 25–52; and the rejoinder by Regina Markell Morantz, "The Perils of Feminist History," *Journal of Interdisciplinary History* 4 (1973): 649–60. All of these essays are included in *Women and Health in America: Historical Readings,* ed. Judith Walzer Leavitt (Madison: University of Wisconsin Press, 1984).

34. L. Emmett Holt, *The Care and Feeding of Children* (New York: D. Appleton, 1894), 51.

35. Julius H. Hess, *Premature and Congenitally Diseased Infants* (Philadelphia: Lea and Febiger, 1922), 27; William A. Silverman, "Neonatal Pediatrics at the Century Mark," *Perspectives in Biology and Medicine* 32 (1989): 162–63.

36. The most concise American review of the incubator temperature issue is provided by John Zahorsky, *Baby Incubators: A Clinical Study of the Premature Infant, with Especial Reference to Incubator Institutions Conducted for Show Purposes* (St. Louis: Courier of Medicine Press, 1905), 41–42.

37. Berthod, 39. Budin did not even mention taking the rectal temperature, focusing instead on sweating, crying, and restlessness as signs of overheating; *Nursling,* 14.

38. V. Hutinel and M. Delestre, "Les couveuses aux Enfants-Assistés," *Revue des maladies de l'enfance* 17 (1899): 534–35.

39. W. A. Silverman, J. W. Fertig, and A. P Berger, "The Influence of the Thermal Environment upon the Survival of Newly Born Premature Infants," *Pediatrics* 22

(1958): 876–90; W. A. Silverman, "Overtreatment of Neonates? A Personal Retrospective," *Pediatrics* 90 (1992): 972.

40. Berthod, 32, 42.

41. On the importance of the mother in supervising the infant see Budin, *Nursling*, 13.

42. Patents were listed annually in U.S. Dept. of the Interior, Patent Office, *Annual Report of the Commissioner of Patents* (Washington, D.C.: Government Printing Office, 1890–95).

43. "An Improved Poultry Brooder," *Scientific American* 66 (16 Apr. 1892): 242; "The Improved 'Monitor' Incubator," *Scientific American* 70 (13 Jan. 1894): 21; "An Electric Incubator," *Scientific American* 73 (21 Dec. 1895): 389.

44. Inventors did have to write reports when they filed patents, and in fact the patent literature for the late 1800s reveals considerable activity in the realm of poultry incubators. In spite of extensive efforts, however, the author has failed to locate any corresponding accounts of infant incubators prior to 1905. Possibly, medical incubators were indexed under a different heading from those used in farming. See annual indexes to U.S. Dept. of Interior, Patent Office, *Annual Reports*, 1890–1905.

45. "Incubator for the Babies: Downy Couch and Warm Air for the Little Ones," *New York Times*, 15 Mar. 1894.

46. Reproduced in Thomas E. Cone, *History of the Care and Feeding of the Premature Infant* (Boston: Little, Brown, 1985), 29.

47. For elaboration of this argument, see Nathan Rosenberg, *Technology and American Economic Growth* (New York: Harper and Row, 1972), 25–31, 57.

48. For accounts of the history of the thermostat, see A. R. J. Ramsey, "The Thermostat or Heat Governor: An Outline of Its History," *Transactions of the Newcomen Society for the Study of the History of Engineering and Technology* 25 (1946): 53–72; F. W. Gibbs, "Invention in Chemical Industries," in *A History of Technology*, ed. Charles Singer, E. J. Holmyard, A. R. Hall, and Trevor I. Williams (Oxford: Clarendon Press, 1957), 3: 679–80.

49. "A Thermostatic Nurse," *Lancet* 1 (1884): 858–59; Ramsey, 58–59.

50. For examples see discussion following "The Rearing of Premature Infants by Means of Incubators," *Pediatrics* 9 (1900): 35; and Zahorsky, 24.

51. For DeLee's biography, see Judith Walzer Leavitt, "Joseph B. DeLee and the Practice of Preventive Obstetrics," *American Journal of Public Health* 78 (1988): 1353–59; Morris Fishbein and Sol Theron DeLee, *Joseph Bolivar DeLee: Crusading Obstetrician* (New York: E. P. Dutton, 1949).

52. DeLee, 25–26.

53. "Inproved 'Monitor' Incubator," 21; "Electric Incubator," 389.

54. A number of examples are cited by Earl V. Haytar, *The Troubled Farmer: Rural Adjustment to Industrialism, 1850–1900* (DeKalb: Northern University Press, 1969), 148.

55. Susan M. Reverby, *Ordered to Care: The Dilemma of American Nursing, 1850–1945* (Cambridge: Cambridge University Press, 1987), 13–16; Ruth Schwartz Cowan, *More Work for Mother: The Ironies of Household Technology from the Open Hearth to the Microwave* (New York: Basic Books, 1983), 119–27.

56. Marx, 313; for a similar assessment, see John A. Lyons, "Incubators and Milk Laboratory Feeding," *American Journal of Obstetrics and Diseases of Women and Children* 36 (1897): 702.

57. "A Thermostatic Nurse," 858; "The Couvreuse *[sic]*, or Mechanical Nurse," 1.

58. For this information on Lion's patent, I am indebted to Dr. Paul Toubas, Children's Hospital of Oklahoma, Oklahoma City, personal communication, 1 Sept. 1994. Lion's own testimony is limited almost exclusively to an 1896 interview with a British journalist: see James Walter Smith, "Baby Incubators," *Strand Magazine* 12 (1896): 770–76.

59. Vallin, "La Maternité Lion de Nice, pour enfants nés avant terme ou débiles, par M. le Dr. Ciaudo," *Bulletin de l'Académie de Médecine*, 3d ser., 34 (1895): 489; "The Use of Incubators for Infants," *Lancet* 1 (1897): 1490.

60. "The Victorian Era Exhibition at Earl's Court," *Lancet* 2 (1897): 161.

61. "Use of Incubators," 1491.

62. Hector Maillart, "Quelques reflexiones sur le functionnement et les resultats des couveuses Lion pendant l'Exposition Nationale," *Revue médicale de la Suisse Romande* 16 (1896): 654.

63. W. Byford Ryan, "The Treatment of Infants Born Prematurely," *Indiana Medical Journal* 8 (1890): 249; Brown, "Care of Premature and Feeble Infants," 160; Marx, 313.

64. Budin described an instance of a premature infant who died when the ward temperature dropped to 50°F (10°C) and its attendant neglected to fill the hot water bottles in its incubator; *Nursling*, 5–6. The older wards were heated by fireplaces or stoves; such dramatic drops of temperature probably did not apply to the *pavillon des débiles*, which featured a steam radiator heating system; *L'assistance publique en 1900* (Paris: Generale de l'assistance publique, 1900), 502. The protection afforded by the walls of the hospital is suggested by Auvard, who wrote in the 1880s that even during the hottest days of July, the ward temperature varied between 65° and 77°F (18–25°C); Auvard, 595.

65. Berthod, 43; Pierre Budin, "Des soins à donner aux nouveau-nés en état de faiblesse congénitale," *La semaine médicale* 8 (1888): 194.

66. "The Couvreuse *[sic]*, or Mechanical Nurse," 334.

67. It was unusual, however, for physicians to mention ventilation specifically with regard to acute symptoms such as suffocation, cyanosis, or respiratory effort. Perhaps this reflected the fact that premature infants were always covered in clothing except for their faces. The principal exceptions were the pediatricians Thomas Rotch and L. Emmett Holt; see Rotch, "Description," 45; L. Emmett Holt, *The Diseases of Infancy and Childhood*, 2d ed. (New York: D. Appleton, 1902), 13.

68. The pediatric leader L. Emmett Holt noted that ventilation became difficult when the ward temperature rose above 75°F, requiring the lid to be cracked open; Holt, *Diseases of Infancy and Childhood*, 13. Holt maintained average ward temperatures in the 66–68°F range; L. Emmett Holt, "The Scope and Limitations of Hospitals for Infants," *Transactions of the American Pediatric Society* 10 (1898): 156.

69. Pierre Budin, "Service des enfants débiles à la Maternité, 1896–1897," *Obstétrique* 4 (1899): 134.

70. Bernard-Jean-Antonin Marfan, *Traité de l'allaitement et de l'alimentation des enfants du premier age* (Paris: G. Steinheil, 1898), quoted in Ernst Deutsch, "Die Lage der Frühgeborenen in den Geburtsanstalten," *Archiv für Kinderheilkunde* 28 (1900): 250.

71. The British statistician and public health reformer William Farr was particularly influential in formulating these ideas; see John W. Eyler, *Victorian Social Medicine: The Ideas and Methods of William Farr* (Baltimore: Johns Hopkins University Press, 1979), 97–107. Florence Nightingale adapted similar reasoning to hospital reform; see Charles Rosenberg, "Florence Nightingale on Contagion: The Hospital as a Moral Universe," in *Healing and History: Essays for George Rosen,* ed. C. E. Rosenberg (New York: Science History Publications, 1979), 116–36.

72. On the history of von Pettenkoffer's theories and their decline, see Charles-Edward Amory Winslow, *Fresh Air and Ventilation* (New York: E. P. Dutton, 1926), 40–46.

73. For a summary of late-nineteenth-century experimental science regarding ventilation, see John Shaw Billings, S. Weir Mitchell, and D. H. Bergey, "The Composition of Expired Air and Its Effects upon Animal Life," *Smithsonian Contributions to Knowledge* 29, no. 989 (1895): 1–13. Billings and his group were focusing on the acute symptoms of suffocation, which they believed to be linked to the sensation of overheating.

74. John Shaw Billings, *Ventilation and Heating* (New York: Engineering Record, 1893), 19–20, 25.

75. Rotch, *Pediatrics,* 309.

76. Rotch, *Pediatrics,* 308–12.

77. Rotch, "Description," 44.

78. Smith, 771–72; "Use of Incubators," 1491.

79. Smith, 773.

80. Florence Nightingale, *Notes on Hospitals* (London: Longman, 1863), facsimile ed., in *Florence Nightingale on Hospital Reform,* ed. Charles Rosenberg (New York: Garland Publishing, 1989), 76.

81. On the role of the fresh air campaigns in the infant mortality crusades of Progressive Era America, see Frank W. Allin, "The Baby Tents of Chicago," *Journal of the American Medical Association* 57 (1911): 2127–28; Robert W. Hastings, "Fresh Air in the Treatment of Children's Diseases," *Pediatrics* 15 (1903): 385–89.

82. Trevor J. Pinch and Wiebe E. Bijker, "The Social Construction of Facts and Artifacts; Or How the Sociology of Science and the Sociology of Technology Might Benefit Each Other," in *The Social Construction of Technological Systems: New Directions in the Sociology and History of Technology,* ed. Wiebe E. Bijker, Thomas P. Hughes, and Trevor J. Pinch (Cambridge: MIT Press, 1987), 40.

5. Propaganada for the Preemies

1. The most comprehensive review of Couney's career is that of William A. Silverman, "Incubator-Baby Side Shows," *Pediatrics* 64 (1979): 127–41; also see L. Joseph Butterfield, "The Incubator Doctor in Denver: A Medical Missing Link," in *The 1970 Denver Westerner's Brand Book* (Denver: Westerner's, 1971), 339–61; Leo Stern, "Ther-

moregulation in the Infant: Historical, Physiological, and Clinical Considerations," in *Historical Review and Recent Advances in Neonatal and Perinatal Medicine,* ed. G. F. Smith, P. N. Smith, and D. Vidyasagar (Chicago: Mead Johnson Nutritional Division, 1983), 1:35–38.

2. The main source of Couney's own testimony is found in the profile by A. J. Liebling, "Patron of the Preemies," *New Yorker 15* (3 June 1939), 20–24. The connection with Budin is also recorded in his obituary in the *New York Times,* 2 Mar. 1950.

3. Vallin, "La Maternité Lion le Nice, pour enfants nés avant terme ou débiles, par M. de Dr. Ciaudo," *Bulletin de l'Académie de Médecine,* 3d ser., 34 (1895): 489.

4. V. Hutinel and M. Delestre, "Les couveuses aux Enfants-Assistés," *Revue des maladies de l'enfance* 17 (1899): 532.

5. Pierre Budin, discussion following M. Fochier, "De l'aération des couveuses par une prise d'air exterieur," *La presse médicale* 14 (1896): clxix; Pierre Budin, *The Nursling: The Feeding and Hygiene of Premature and Full-Term Infants,* trans. William J. Maloney (London: Caxton Publishing, 1907), 13.

6. The cost of the Lion incubator in American dollars was around $150 in the early 1900s; William H. Mercur, to Joseph B. DeLee, Pittsburgh, 19 Dec. 1902, DeLee Papers, Northwestern Memorial Hospital Archives, Chicago.

7. Vallin, 489.

8. Lion's incubator charity was one of a number of French institutions appearing in the 1890s directed at improving infant health outside the hospital. Most of these centered on the prevention of diarrheal disease, the leading cause of death in infancy. Milk depots, or *gouttes de lait,* chiefly provided sterile milk, while Budin's *consultations des nourrissons* promoted breast-feeding through physician supervision and weight checks. Like these other institutions, the Maternité Lion represented an attempt to find alternatives to the hospital to promote infant health. Alisa Klaus, *Every Child a Lion: The Origins of Maternal and Infant Health Policy in the United States and France, 1890–1920* (Ithaca: Cornell University Press, 1993), 62–67.

9. James Walter Smith, "Baby Incubators," *Strand Magazine* 12 (1896): 775–76.

10. Vallin, 490. This report summarizes Ciaudo's study, presenting the statistics on 489–90.

11. Smith, 770–71; "Paris Letter: An Improved System of Incubators," *Pediatrics* 1 (1896): 427.

12. Smith, 775–76. Qualifying these comments is the paucity of information on wet nurses in Lion's establishments, who, though apparently present to provide human milk, were not mentioned explicitly in Smith's account.

13. Liebling, 20–22.

14. Ibid.

15. "The Victorian Era Exhibition at Earl's Court," *Lancet* 2 (1897): 161; Liebling, 20.

16. "An Artificial Foster-Mother: Baby Incubators in the Berlin Exposition," *Graphic,* 10 Oct. 1896, 461; Smith, 776. Lion was actively promoting his incubator charities beyond France at the time, recruiting physicians from Geneva and Berlin; Hector Maillart, "Quelques reflexiones sur le fonctionnement et les resultats des couveuses Lion pendant l'Exposition Nationale," *Revue médicale de la Suisse Romande* 16 (1896): 644.

17. "The Danger of Making a Public Show of Incubators for Babies," *Lancet* 1 (1898): 390–91; for Couney's defense, see Samuel Schenkein and Martin Coney [*sic*], "Infant Incubators," *Lancet* 2 (1897): 744.

18. On the Altmann model, see "Victorian Era Exhibition," 161; and "Letter from Buffalo," *New York Medical Journal* 73 (1901): 103; on Kny-Scheerer, see Joseph B. DeLee, "Infant Incubation, with the Presentation of a New Incubator and a Description of the System at the Chicago Lying-in Hospital," *Chicago Medical Recorder* 22 (1902): 24.

19. "The Use of Incubators for Infants," *Lancet* 1 (1897): 1490–91; also see "Victorian Era Exhibition," 161. Incubator manufacturers did in fact rent their products in London, removing some of the financial obstacles to their use. See James Frederic Goodhart, *The Diseases of Children*, 8th ed., ed. George Frederic Still (London: Churchill, 1905), 28.

20. "Danger of Making a Public Show," 390–91.

21. "Incubators in London," *Pediatrics* 5 (1898): 298–99.

22. A brief description is found in the *Official Guidebook to the Trans-Mississippi and International Exposition* (Omaha: Megeath Stationary Company, 1898; Smithsonian Books of the Fairs, microfilm), 48. Couney later claimed that he transported the infants used in this show by train from Chicago; I have found no other accounts to confirm this (Liebling, 22).

23. Liebling, 20. An advertisement for the incubator exhibit at the Paris exposition made clear that it was managed by Lion and offers further testimony to his linkage with Couney; William A. Silverman, M.D., personal collection, Greenbrae, Calif.

24. "Exhibit of Infant Incubators at the Pan-American Exposition," *Pediatrics* 12 (1901): 414–19; "Baby Incubators at the Pan-American Exposition," *Scientific American* 85 (3 Aug. 1901): 68; Arthur Brisbane, "The Incubator Baby and Niagara Falls," *Cosmopolitan* 31 (1901): 509–16.

25. Butterfield, 351–52; Liebling 20.

26. Quoted in Robert W. Rydell, *All the World's a Fair: Visions of Empire at American International Expositions, 1876–1916* (Chicago: University of Chicago Press, 1984), 4.

27. Reid Badger, *The Great American Fair: The World's Columbian Exposition and American Culture* (Chicago: Nelson-Hall, 1979), 17.

28. "La Couveuse, for Preserving the Life of Feeble and Prematurely Born Infants," *Medical Record* 37 (1890): 721; Thomas M. Rotch, "Description of a New Incubator," *Transactions of the American Pediatric Society* 5 (1893): 44–46. On the role of expositions in technology transfer, see Thomas P. Hughes, *Networks of Power: Electrification in Western Society, 1880–1930* (Baltimore: Johns Hopkins University Press, 1983), 50–51.

29. Badger, 18–19.

30. Paul Greenhalgh, *Ephemeral Vistas: The Expositions Universelles, Great Expositions, and World's Fairs, 1851–1939* (Manchester, U.K.: Manchester University Press, 1988), 86–87.

31. Robert Bogdan, *Freak Show: Presenting Human Oddities for Amusement and Profit* (Chicago: University of Chicago Press, 1988), 2, 35–46.

32. Ibid., 289.

33. H. C. Allen, "Early Viability of Twin Fetuses," *New York Medical Journal* 58 (1893): 208.

34. "Exhibit of Infant Incubators at the Pan-American Exhibition," *Pediatrics* 12 (1901): 414–19; "Letter from Buffalo," 1038–40.

35. Robert Grant, "Notes on the Pan-American Exposition," *Cosmopolitan* 31 (1901): 462. See map in Mark Bennett, *The Pan-American Exposition and How to See It* (Buffalo: Goff, 1901).

36. Butterfield, 351–52; John Zahorsky, *Baby Incubators: A Clinical Study of the Premature Infant* (St. Louis: Courier of Medicine Press, 1905), 14–18.

37. Liebling, 20.

38. "Qbata Infant Incubators at the Pan-American Exposition," *National Magazine* 14 (1901): 551.

39. *World's Fair Bulletin*, Apr. 1904, quoted in Butterfield, 352.

40. "Letter from Buffalo," 1040; "Exhibit of Infant Incubators," 419. The origins of the 85 percent survival rate figure are not clear but may derive from adjusting Lion's statistics by subtracting early deaths or very young infants. By the time of the 1897 London exposition, Couney appears to have claimed an overall survival rate of 80–90 percent; "Incubators in London," 298.

41. "Qbata Incubators," 554.

42. "Artificial Foster-Mother", 461.

43. Brisbane, 554.

44. Ibid., 515–16. It is interesting that many popular accounts of Couney's exhibits (including Brisbane's) claimed that premature infants were both deaf and blind, a belief that would tend to undercut the mother's interaction and help remove some of the objections to raising such infants on the Midway. "Artificial Foster-Mother," 461; Brisbane, 516.

45. John F. Kasson, *Amusing the Million: Coney Island at the Turn of the Century* (New York: Hill and Wang, 1978), 9.

46. Bogdan, 56; Kasson, 50.

47. Edo McCullough, *Good Old Coney Island: A Sentimental Journey into the Past* (New York: Charles Scribner's Sons, 1957), 276.

48. Liebling, 20.

49. McCullough, 276, 279.

50. McCullough, 277.

51. *Journal of the Senate of the State of New York, 129th Session* (Albany: Brandow Printing, 1906), 1:16. I have been unable to locate specific hearings.

52. "Flames Sweep Coney Island; Incubator Babies Killed," *New York Times*, 27 May 1911. Subsequent accounts asserted that the supervising physician at the Dreamland exhibit, Dr. J. Fischel, and his nurses rescued the infants and transported them to a New York hospital. Couney was not mentioned and may have been at the Luna Park exhibit. See "Start Up Again in Coney Ruins; Incubator Babies Safe," *New York Times*, 28 May 1911; and "Failure of Workmen to Give Alarm Let Fire Spread Beyond Control," *New York Herald*, 28 May 1911.

53. "John Douglas Lindsay," *National Cyclopedia of American Biography* (New York: James T. White, 1944), 26: 191–92. Though Lindsay's background may suggest

sympathy with the powerful antivivisection movement of his day, his 1910 letter (cited below) to the *New York Times* fell short of linking the shows to a more general condemnation of the medical profession and specifically asserted that premature infant care belonged in the hospital under the direction of physicians. On antivivisectionists as child advocates, see Susan Eyrich Lederer, "Hideyo Noguchi's Luetin Experiment and the Antivivisectionists," *Isis* 76 (1985): 31–48.

54. John D. Lindsay to editor, 27 May 1911, reprinted in "Start Up Again." Lindsay's statement was also cited in "Babies in Incubator Endangered by Fire," *New York Herald*, 28 May 1911.

55. Liebling, 21.

56. Silverman, 133.

57. De Witt H. Sherman, "The Premature Infant," *New York Medical Journal* 82 (1905): 274–75.

58. John Zahorsky, *From the Hills: An Autobiography of a Pediatrician* (St. Louis: C. V. Mosby, 1949), 15–26, 154–58, 183–84.

59. John Zahorsky, *Baby Incubators: A Clinical Study of the Premature Infant, with Especial Reference to Incubator Institutions Conducted for Show Purposes* (St. Louis: Courier of Medicine Press, 1905), 14–15.

60. Ibid.

61. Julius H. Hess, *Premature and Congenitally Diseased Infants* (Philadelphia: Lea and Febiger, 1922).

62. Zahorsky, *Baby Incubators*, 12.

63. Ibid., 13.

64. Ibid., 14–18, 37–39.

65. Ibid., 39–52.

66. Ibid., 56–85.

67. Exemplifying the interest in premature infant nutrition was the Boston pediatrician John Lovett Morse; see his "A Study of the Caloric Needs of Premature Infants," *American Journal of the Medical Sciences* 127 (1904): 463–77. The caloric approach to feeding was rooted in the metabolic studies of the German researchers Max Rubner and Otto Heubner in 1898; see Richard A. Meckel, *Save the Babies: American Public Health Reform and the Prevention of Infant Mortality, 1850–1929* (Baltimore: Johns Hopkins University Press, 1990), 59–60.

68. Zahorsky, *From the Hills*, 166.

69. Dr. Tuttle, quoted in Vitray Papin Blair, "Premature Infants: The Necessity for and the Difficulty of Formulating a General Plan for Their Care," *St. Louis Courier of Medicine* 33 (1905): 246.

70. A very brief, though favorable, review appeared in *American Journal of Obstetrics and Diseases of Women and Children* 53 (1906): 586. No reviews appeared between 1905 and 1907 in *Archives of Pediatrics* or *Pediatrics*.

71. Harold Kniest Faber and Rustin McIntosh, *History of the American Pediatric Society, 1887–1965* (New York: McGraw-Hill, 1966), 12, 51–52; Zahorsky, *From the Hills*, 183–84.

72. Zahorsky, *Baby Incubators*, 133. By 1914, Zahorsky advocated employing a simple homemade incubator made out of a soap box; see comment in H. M. McClana-

han, "Management of Delicate and Premature Infants in the Home," *Journal of the American Medical Association* 63 (1914): 1759.

73. Zahorsky, *From the Hills,* 170, 181–82, 294–301.

74. Zahorsky, *Baby Incubators,* 28–31; Zahorsky, *From the Hills,* 165, 181.

75. Zahorsky, *Baby Incubators,* 14.

76. Zahorsky, *From the Hills,* 167.

77. Lindsay to editor.

78. Couney's major exhibits after the Buffalo Pan-American Exposition of 1903 were Portland (1905), Mexico City (1908), Rio de Janeiro (1910), and San Francisco (1915), followed by a eighteen-year hiatus until the appearance at the Chicago Century of Progress Exposition in 1933. His appearances at amusement parks are more difficult to trace but included Denver (1912) and Chicago (1914). Finally, he maintained his exhibits at Coney Island and the Atlantic City boardwalk from 1903 to the Second World War. The major exhibits were listed in a letter, Martin Couney to Julius Hess, 18 October 1940, Julius H. Hess Papers, Regenstein Library, University of Chicago. The sequence has been reconstructed in Silverman, 127–41; on the Denver and Chicago shows, see Butterfield, 339–61.

79. Kasson, 112.

80. The only point between 1910 and 1930 at which Couney came to the attention of the *New York Times* was a 1926 incident in which he was charged with striking a policeman with his auto to evade a parking violation. "Doctor Denies That He Hit Policeman with Auto," *New York Times,* 24 July 1926. An investigation later cleared Couney of the charges; *New York Times,* 4 Aug. 1926.

81. Henry D. Chapin, discussion following Walter Lester Carr, "A Clinical Report of Simple Methods in the Care of Premature Babies," *Archives of Pediatrics* 38 (1921): 402–3; John A. Foote, *Diseases of the New-born: A Monographic Handbook* (Philadelphia: J. B. Lippincott, 1926), 31.

82. Joseph S. Wall, "The Status of the Child in Obstetric Practice," *Journal of the American Medical Association* 66 (1916): 256.

83. On the remarkable proliferation of electrotherapeutic devices in the late 1800s, see L. Rosner, "The Professional Context of Electrotherapeutics," *Journal of the History of Medicine and Allied Sciences* 43 (1988): 64–82; and Lawrence D. Longo, "Electrotherapy in Gynecology: The American Experience," *Bulletin of the History of Medicine* 60 (1986): 343–66.

84. Accounts of the meeting are more consistent on the timing (1914) than on exact details. Obituaries by Morris Fishbein and Hess' successor Ralph Kundstadter (*Chicago Medical Society Bulletin,* 26 January 1957) are cited in Butterfield, 358. William Silverman has collected other stories, mostly conflicting; 136.

85. In 1907, Hess' hospital, Michael Reese, set aside two rooms to be developed into an incubator station; at that point incubators were already in operation in the main nursery; see *All Our Lives: A Centennial History of Michael Reese Hospital and Medical Center, 1881–1981,* ed. Sarah Gordon (Chicago: Michael Reese, 1981), 61, 74. Hess's first article on premature infants was Julius H. Hess, "A Study of the Caloric Needs of Premature Infants," *American Journal of Diseases of Children* 2 (1911): 302–14. On the incubator, see Julius H. Hess, "An Electric-Heated Water-Jacketed Infant In-

cubator and Bed, for Use in the Case of Premature and Poorly Nourished Infants," *Journal of the American Medical Association* 64 (1915): 1068–69.

86. Julius H. Hess, *Premature and Congenitally Diseased Infants,* vi.

87. Liebling, 20–24; Silverman, 127–41.

88. John Harley Warner, "Science in Medicine," *Osiris,* 2d ser., 1 (1985): 37–58; John Harley Warner, "Ideals of Science and Their Discontents in Late-Nineteenth-Century American Medicine," *Isis* 82 (1991): 454–78.

89. Kenneth M. Ludmerer, *Learning to Heal: The Development of American Medical Education* (New York: Basic Books, 1985), 47–190.

90. Richard Hofstadter, *The Progressive Historians: Turner, Beard, Parrington* (New York: Knopf, 1968), 37.

91. Stanley Joel Reiser, *Medicine and the Reign of Technology* (Cambridge: Cambridge University Press, 1978), 91–121.

6. The Experiment in Obstetric Neonatology

1. M. Howard Fussell, "Obstetrics and the General Practitioner," *Journal of the American Medical Association* 39 (1902): 1629.

2. Sydney A. Halpern, *American Pediatrics: The Social Dynamics of Professionalism, 1880–1980* (Berkeley: University of California Press, 1988), 35–56.

3. Susan M. Reverby, *Ordered to Care: The Dilemma of American Nursing, 1850–1945* (Cambridge: Cambridge University Press, 1987), 14–15.

4. William H. Taylor, "Some Points in Relation to Premature Children," *American Journal of Obstetrics and the Diseases of Women and Children* 20 (1887): 1022.

5. In 1887, for example, one physician writing an article on premature infant care was unaware of Tarnier's invention, recommending instead warming the infant near an open fireplace: Taylor, 1025. Advocates of the incubator in the late 1890s often called attention to its relative unpopularity as compared with that in Europe. See L. Emmett Holt, *The Diseases of Infancy and Childhood* (New York: D. Appleton, 1897), 14; John A. Lyons "Incubators and Milk Laboratory Feeding," *American Journal of Obstetrics and Diseases of Women and Children* 36 (1897): 696.

6. Henry Bettmann, "Premature Labor and the New-Born Child," *American Journal of Obstetrics and Diseases of Women and Children* 25 (1892): 324–25; Pierre Budin and M. Perret, "Nouvelle recherches sur les enfants débiles," *Obstétrique* 6 (1901): 215–19.

7. Bettmann, 324.

8. R. B. Gilbert, "The Care of Premature Infants after Induced and Accidental Labors," *Transactions of the Kentucky Medical Society,* n.s., 5 (1896): 189–95.

9. James D. Voorhees, "The Care of Premature Babies in Incubators," *Archives of Pediatrics* 17 (1900): 331–36; James D. Voorhees, "Dilatation of the Cervix by Means of a Modified Champetier des Ribes Balloon," *Medical Record* 58 (1900): 361–66.

10. W. Byford Ryan, "The Treatment of Infants Born Prematurely," *Indiana Medical Journal* 8 (1890): 249.

11. Ibid., 248–49.

12. S. Marx, "Incubation and Incubators," *American Medico-Surgical Bulletin* 9 (1896): 311.

13. Henry J. Garrigues, discussion following Robert P. Harris, "The Present and Improving Status of Caesarean Surgery," *Transactions of the American Gynecological Society* 16 (1891): 135.

14. S. W. Ransom, "The Care of Premature and Feeble Infants," *Pediatrics* 9 (1900): 322.

15. On the history of midwives in the United States after the mid-nineteenth century, see Judy Barrett Litoff, *American Midwives: 1860 to the Present* (Westport, Conn.: Greenwood Press, 1978); and Frances E. Kobrin, "The American Midwife Controversy: A Crisis of Professionalization," *Bulletin of the History of Medicine* 40 (1966): 350–63.

16. On the rise of surgery in American hospitals, see Charles E. Rosenberg, *The Care of Strangers: The Rise of America's Hospital System* (New York: Basic Books, 1987), 147–50.

17. Harold Speert, *Obstetrics and Gynecology in America: A History* (Baltimore: Waverly Press, 1980), 119–21.

18. "Letter from New York," *Journal of the American Medical Association* 6 (1886): 697–99; A. H. Halberstadt, "Advances in Obstetrics during Last Half Century," *Journal of the American Medical Association* 36 (1901): 1168; F. Winckel, "The Necessity of the Union of Obstetrics and Gynecology as Branches of Medical Instruction," *Transactions of the American Gynecological Society* 18 (1893): 18–35.

19. Richard A. Meckel, *Save the Babies: American Public Health Reform and the Prevention of Infant Mortality, 1850–1929* (Baltimore: Johns Hopkins University Press, 1990), 102–3; Alisa Klaus, *Every Child a Lion: The Origins of Maternal and Infant Health Policy in the United States and France, 1890–1920* (Ithaca: Cornell University Press, 1993), 10–42.

20. Kenneth M. Ludmerer, *Learning to Heal: The Development of American Medical Education* (New York: Basic Books, 1985), 29–38; Thomas Bonner, *American Doctors and German Universities: A Chapter in International Intellectual Relations, 1870–1914* (Lincoln: University of Nebraska, 1963).

21. "French Medical Journals," *Medical Record* 24 (1883): 545.

22. James R. Chadwick, "Obstetric and Gynaecological Literature, 1876–1881," *Boston Medical and Surgical Journal* 105 (1881): 247.

23. Ibid., 245–46. Chadwick noted that the French tended to produce a higher proportion of books and theses than did physicians from other countries. He speculated that Parisian medical education and hospital rivalry encouraged a theoretical preoccupation among its academic leaders.

24. John Harley Warner, *The Therapeutic Perspective: Medical Practice, Knowledge, and Identity in America, 1820–1885* (Cambridge: Harvard University Press, 1986), 274, 280.

25. "The Place of Ergot in Obstetric Practice," *Journal of the American Medical Association* 21 (1893): 543; Fussell, 1629–32; Helen Hughes, "A Consideration of Some Points in Obstetrics," *New York Medical Journal* 80 (1904): 202–5.

26. Henry P. Newman, "Section of Obstetrics and Diseases of Women: Address of Chairman Delivered at the Fifty-Second Annual Meeting of the AMA," *Journal of the American Medical Association* 36 (1901): 1758; E. Gustav Zinke, "The Practice of Obstetrics," *Journal of the American Medical Association* 37 (1901): 610.

27. Edward Reynolds, "Circumstances Which Render the Elective Section Justifiable in the Interests of the Child Alone," *Transactions of the American Gynecological Society* 26 (1901): 277–82; E. Gustav Zinke, "The Limitations of Cesarean Section," *Transactions of the American Association of Obstetrics and Gynecology* 16 (1903): 29–36.

28. See Gert H. Brieger, "A Portrait of Surgery: Surgery in America, 1875–1889," *Surgical Clinics of North America* 67 (1987): 1181–16.

29. On nineteenth-century maternity hospitals, see Judith Walzer Leavitt, *Brought to Bed: Childbearing in America, 1750–1950* (New York: Oxford University Press, 1986), 74; Rosenberg, 269–70; Morris J. Vogel, *The Invention of the Modern Hospital: Boston, 1870–1930* (Chicago: University of Chicago Press, 1980), 12–13.

30. Zinke, "Practice of Obstetrics," 610–13.

31. Among these were substantial new buildings built by the Philadelphia Lying-in Charity (1888), New York's Sloane Maternity Hospital (1899), the New York Lying-in Hospital (1902), the Woman's Hospital of Philadelphia (1895, 1908), and the Women's Hospital of the State of New York (1906–15). See Speert, *Obstetrics and Gynecology,* 91–102.

32. Leavitt, *Brought to Bed,* 177; Vogel, 117.

33. Adelaide Brown, "Care of Premature and Feeble Infants," *Transactions of the Medical Society of California* 26 (1896): 156.

34. Vilray P. Blair, "Some Notes on the Care of Premature Infants," *Pediatrics* 16 (1904): 594.

35. Joseph B. DeLee, "The Prophylactic Forceps Operation," *American Journal of Obstetrics and Gynecology* 1 (1920): 34–35, 39–41.

36. For an insightful interpretation of the prophylactic forceps operation as a medical technology presented in the language of preventive medicine see Judith Walzer Leavitt, "Joseph B. DeLee and the Practice of Preventive Obstetrics," *American Journal of Public Health* 78 (1988): 1353–59.

37. For DeLee's biography, see Morris Fishbein and Sol Theron DeLee, *Joseph Bolivar DeLee: Crusading Obstetrician* (New York: E. P. Dutton, 1949); and Leavitt, "Joseph B. DeLee," 1353–59.

38. Morris Fishbein, abstract of interview with Dr. Isaac Abt, 26 Sept. 1945, Joseph B. DeLee Papers, Northwestern Memorial Hospital Archives, Chicago.

39. Joseph B. DeLee, "A Very Brief History of the Chicago Lying-in Hospital and Dispensary," pamphlet, 1931, DeLee Papers, 5–11. On Addams's involvement, see "Seek to Oust Hospital: Twentieth-Ward Residents Open War on Small Places," newspaper clipping from *Chicago Chronicle,* 29 June 1903; "Off Nuisance List: Health Committee of Council Frees Lying-in Hospitals from Restrictions," unspecified newspaper clipping, 23 May 1903, DeLee Papers.

40. Chicago Lying-in Hospital and Dispensary, *Annual Reports* (Chicago: S. Ettlinger Printing, 1899–1900); 7, (henceforth ChLy, *AR*). ChLy, *AR* (1900–1901), 6; ChLy, *AR* (1901–2); 15; DeLee papers. Also available in Archives of Chicago Historical Society, Chicago.

41. Joseph B. DeLee, "Chicago Lying-in Hospital and Dispensary," reprint from *Neoplasm* (Northwestern Medical School Yearbook, 1903), DeLee Papers.

42. DeLee visited many of the Paris maternity hospitals during his postgraduate European studies in 1894, according to the recollection of a travel companion; Walter L. Bierring to Morris Fishbein, 16 October 1945, DeLee Papers.

43. Joseph B. DeLee, "Infant Incubation, with the Presentation of a New Incubator and a Description of the System at the Chicago Lying-in Hospital," *Chicago Medical Recorder* 22 (1902): 23.

44. DeLee, "Infant Incubation," 26. DeLee demonstrated his knack for technical gadgets as early as high school, when he earned extra money installing electric doorbells and wiring gas fixtures. Isaac A. Abt, *Baby Doctor* (New York: Whittlesey House, 1944), 105; Fishbein, 36.

45. For an extensive discussion of these technical issues, see DeLee, "Infant Incubation," 25–30. DeLee was not alone in having trouble with the ventilation of Lion incubators. John Zahorsky, the pediatrician supervising a dozen Lion incubators at the Louisiana Purchase Exposition in 1904, had great difficulty with the system, noting that airflow almost stopped at times during the summer. The filter to remove bacteria could also interfere with airflow. Zahorsky, *Baby Incubators: A Clinical Study of the Premature Infant, with Especial Reference to Incubator Institutions Conducted for Show Purposes* (St. Louis: Courier of Medicine Press, 1905), 24–26.

46. DeLee published his article "Infant Incubation" in the following journals: *Bulletin of Northwestern University Medical School* 5 (1903–4): 252–64; *National Hospital Record* 7 (1903–4): 10–17; *British Journal of Nursing* 32 (1904): 387–90; as well as the *Chicago Medical Recorder* 22 (1902): 22–40. All citations in this paper refer to the latter. The thermostat was the main innovation that enabled DeLee to market the incubator as his own, at a cost thirty dollars higher than the Kny-Scheerer price of $150; William H. Mercur to Joseph B. DeLee, Pittsburgh, 19 Dec. 1902, DeLee Papers. Though he clearly sought to patent his incubator [see ChLy, *AR* (1902–3), 16; ChLy, *AR* (1901–2), 15], no patents are indicated between 1902 and 1908 in U.S. Dept. of the Interior, Patent Office, *Annual Report of the Commissioner of Patents* (Washington, D.C., Government Printing Office, 1902–8). Eventually, the Scanlan-Morris Company of Madison, Wisconsin, manufactured his model, according to the catalogue of its distributor V. Mueller and Company (Chicago, Illinois, 1925); DeLee Papers. Scanlan-Morris later manufactured Julius Hess's incubator in the 1920s and 1930s.

47. DeLee, "Infant Incubation," 30.

48. For a succinct introduction to the concept of technological systems, see Thomas P. Hughes, "The Evolution of Large Technological Systems," in *The Social Construction of Technological Systems: New Directions in the Sociology and History of Technology*, ed. Wiebe E. Bijker, Thomas P. Hughes, and Trevor J. Pinch (Cambridge: MIT Press, 1987), 51–82. Also see Thomas P. Hughes, *American Genesis: A Century of Invention and Technological Enthusiasm, 1870–1970* (New York: Viking Press, 1989), 1; and Thomas P. Hughes, *Networks of Power: Electrification in Western Society, 1880–1930* (Baltimore: Johns Hopkins University Press, 1983), 18–21.

49. Hughes, *Networks of Power*, 18–21.

50. Hughes, *Networks of Power*, 78.

51. DeLee, "Infant Incubation," 39.

52. In his 1903–4 annual report, DeLee excluded the deaths of eleven "moribund"

infants from his total of twenty-six admissions and thirteen deaths in presenting an overall survival rate that amounted to 87 percent instead of 50 percent; ChLy, *AR* (1903–4), 29. In other years this factor made less of a difference. In 1905, he reported that since its creation the station had received 80 viable infants from a total of 105, of which 70 lived; DeLee, "The Chicago Lying-In Hospital and Dispensary," *Reform Advocate*, 7 Jan, 1905, 464, DeLee Papers.

53. Joseph B. DeLee, *Obstetrics for Nurses*, 3d ed. (Philadelphia: Saunders, 1909), 378; DeLee, "Chicago Lying-In Dispensary and Hospital," clipping from *Reform Advocate*, about 1902, 78–81, DeLee Papers; ChLy, *AR* (1899–1900), 7.

54. "Smallest Ambulance in the World," newspaper clipping, 12 May 1903, DeLee Papers.

55. DeLee, *Obstetrics for Nurses*, 374.

56. ChLy, *AR* (1903–4), 29; 1 ChLy, *AR* (1904–5), 30; ChLy, *AR* (1905–6), 28; ChLy, *AR* (1907–8), 37.

57. DeLee, "Infant Incubation," 38.

58. Joseph B. DeLee, "Asphyxia Neonatorum: Causation and Treatment," *Medicine* 3 (1897): 643–50.

59. "Report of the Incubator Station," ChLy, *AR* (1901–02), 15.

60. DeLee, "Asphyxia Neonatorum," 643–50.

61. DeLee, "Very Brief History," 9, 12.

62. DeLee, *Obstetrics for Nurses*, 383.

63. E. E. Koch, "The Immediate Care of a Premature Child," *American Journal of Nursing* 6 (1905–6): 509.

64. Reverby, 143–58. On scientific management and hospital organization in the early 1900s, see Edward T. Mormon, introduction to *Efficiency, Scientific Management, and Hospital Standardization: An Anthology of Sources* (New York: Garland Publishing, 1989).

65. Abt, 105.

66. Koch, 508.

67. Fishbein, 70.

68. ChLy, *AR* (1904–5), 24, 30.

69. On fund-raising and the station's expansion, see ChLy, *AR* (1901–2), 17; ChLy, *AR* (1903–4), 6. Regarding DeLee's contribution, see an envelope of paid bills for equipment and hospital expenses mostly for incubator materials, paid by DeLee and totaling $1,094.50; DeLee Papers.

70. ChLy, *AR* (1907–8), 27.

71. See photographs of Lion incubator on baby ward in ChLy, *AR* (1908–10), 15. The annual reports revealed only six to eight annual premature infant admissions in 1908–10 and only two to six a year between 1911 and 1914; see ChLy, *AR* (1911–14): 52–53.

72. Joseph B. DeLee, "The Chicago Lying-in Hospital and Dispensary," *Modern Hospital* 4 (1915): 387.

73. On hospital economics in the early twentieth century, see Rosemary Stevens, *In Sickness and in Wealth: American Hospitals in the Twentieth Century* (New York: Basic Books, 1989), 17–51.

74. Examples include Charles Herrmann, discussion following L. Emmett Holt

and Ellen C. Babbit, "Institutional Mortality of the Newborn: A Report on Ten Thousand Consecutive Births at the Sloane Hospital for Women, New York," *Transactions of the American Association for the Study and Prevention of Infant Mortality* 5 (1914): 164; Joseph S. Wall, "The Status of the Child in Obstetric Practice," *Journal of the American Medical Association* 66 (1916): 255–56. On formal pediatric efforts to gain access to newborns, see Halpern, 71.

75. Lyons, 699.

76. See the discussions appended to the following articles: Vilray Papin Blair, "An Incubator for Delicate or Premature Infants," *Medical Fortnightly* 23 (1903): 468–71; Vilray Papin Blair, "Premature Infants: The Necessity for and the Difficulty of Formulating a General Plan for Their Care," *St. Louis Courier of Medicine* 33 (1905): 240–49; "The Rearing of Premature Babies by Means of Incubators," *Pediatrics* 9 (1900): 35–36; and John Lovett Morse, "The Care and Feeding of Premature Babies," *American Journal of Obstetrics and Diseases of Women and Children* 51 (1905): 687–88. On incubator treatment at Boston Lying-in Hospital, see John Lovett Morse, "A Study of the Caloric Needs of Premature Infants," *American Journal of the Medical Sciences* 127 (1904): 467.

77. Francis H. Stuart, "De Lion Incubator at Low Maternity Hospital," *Brooklyn Medical Journal* 15 (1901): 346, 349.

78. Stevens, 106–14.

79. Voorhees, "The Care of Premature Babies," 332–36. On incubators in the Sloane nursery, see Harold Speert, *The Sloane Hospital Chronicle: A History of the Department of Obstetrics and Gynecology of the Columbia-Presbyterian Medical Center* (Philadelphia: F. A. Davis, 1963), 133.

80. Sloane began to accept large numbers of private patients for the first time in its history between 1897 and 1899, during which time it moved to a new six-story structure as well; Rosenberg, 404 n. 29; Speert, *Sloane Hospital Chronicle*, 124–26. On Sloane's early history, also see T. Gaillard Thomas, "Address at the Inauguration of the Sloane Maternity Hospital and the Vanderbilt Clinic, December 29, 1887," *New York Medical Journal* 47 (1888): 31.

81. Voorhees also used statistics to assess his premature-labor-induction balloon; see Voorhees, "Dilatation of the Cervix," 361–66. For Voorhees's biography, see entry in *National Cyclopedia of American Biography* (New York: James T. White, 1944), 31:151–52; Speert, *Obstetrics and Gynecology*, 227.

82. Henry Dwight Chapin and Godfrey Roger Pisek, *Diseases of Infants and Children* (New York: William Wood, 1909), 1; L. Emmett Holt, *The Diseases of Infancy and Childhood*, 2d ed. (New York: D. Appleton, 1902), 14; "Electric Light Saving the Babies," *Literary Digest* 89 (8 May 1926): 22.

83. The reasons for this difference are hard to judge in retrospect, particularly since the comparability of the two populations remains uncertain. Neither Tarnier nor Voorhees employed incubators for all their premature infants but allocated them to as many as possible based on availability. This factor may explain why the mortality rates for the eight-month infants differed far more than those of seven-month infants; Voorhees may well have had to reserve his incubators for the smaller babies to a greater extent than had Tarnier. A second factor might also explain why Voorhees's

mortality rates were higher overall. Unlike the Maternité under Tarnier, Sloane had to transfer a number of incubator babies to infant hospitals, where they faced high mortality rates. Voorhees included these infants, thereby raising his overall mortality rate from roughly 40 percent to 50 percent. Voorhees, "Care of Premature Babies," 332, 342–45.

84. Ibid., 341.

85. Edward A. Ayers, discussion following Chapin, 36.

86. The pediatric leader Abraham Jacobi believed that midwives were more likely than obstetricians to care for the premature infant; Jacobi, "The Best Means of Combating Infant Mortality," *Journal of the American Medical Association* 58 (1912): 1737.

87. The principal pediatricians cited were both from Boston: Thomas Morgan Rotch (see Lyons, 698; and Adelaide Brown, 161) and John Lovett Morse [see Jennings C. Litzenberg, "The Care of Premature Infants, with Special Reference to the Use of Home-Made Incubators," *Journal of the Minnesota Medical Association* 28 (1908): 88.] Morse also gave an address to the Washington Obstetric and Gynecologic Society in 1905; Morse, "Care and Feeding of Premature Infants," 589–99.

88. Mary Breckinridge, *Wide Neighborhoods: A Story of the Frontier Nursing Service* (New York: Harper and Brothers, 1952), 56.

89. Speert, *Obstetrics and Gynecology*, 92–94.

90. Voorhees, "Care of Premature Babies," 341, 342–45. Another example was provided by John Lovett Morse of Boston, who reported a series of six premature babies in 1904, three of whom had been transferred after two weeks of incubator care in the Boston Lying-in Hospital; see Morse, "Study of the Caloric Needs of Premature Infants," 467.

91. L. Joseph Butterfield, "Historical Perspectives of Neonatal Transport," *Pediatric Clinics of North America* 40 (Apr. 1993): 221–39.

92. Zahorsky, 37–38. The inspiration behind Zahorsky's transport incubator is uncertain but may well have been DeLee, given that Couney played down the importance of transport, claiming to have relied on padded baskets; A. J. Liebling, "Patron of the Preemies," *New Yorker* 15 (3 June 1939), 22. At least some St. Louis physicians had personally seen DeLee's transport incubator at the same time; F. J. Taussig, discussion following Vilray Papin Blair, "Premature Infants: The Necessity for and the Difficulty of Formulating a General Plan for Their Care," *St. Louis Courier of Medicine* 33 (1905): 244.

93. Julius H. Hess, "Heated Bed for Transportation of Premature Infants," *Journal of the American Medical Association* 80 (1923): 1313; Thomas E. Cone, Jr., *History of the Care and Feeding of the Premature Infant* (Boston: Little, Brown, 1985), 80–81; Butterfield, 232–33.

94. Louis A. Shaw and Philip Drinker, "An Apparatus for the Prolonged Administration of Artificial Respiration," *Journal of Clinical Investigation* 8 (1929): 33; Cone, 122–29.

7. The Pediatric Revolt

1. Harold Kniest Faber and Rustin McIntosh, *History of the American Pediatric Society 1887–1965* (New York: McGraw-Hill, 1966), 12, 17–55.

2. Thomas M. Rotch, *Pediatrics: The Hygienic and Medical Treatment of Children* (Philadelphia: J. B. Lippincott, 1895), 288–317. For comparison, the leading American pediatric textbook of the time devoted only five pages to the premature infant; L. Emmett Holt, *The Diseases of Infancy and Childhood* (New York: D. Appleton, 1897), 10–14.

3. For Rotch's biography, see Fritz B. Talbot, "Thomas M. Rotch," in *Pediatric Profiles*, ed. Borden Smith Veeder (St. Louis: C. V. Mosby, 1957), 29–32; Thomas E. Cone, Jr., "Thomas Morgan Rotch," in *Dictionary of American Biography*, ed. Martin Kaufman, Stuart Galishoff, and Todd Savitt (Westport, Conn.: Greenwood Press, 1984), 2:650–51.

4. On Rotch's feeding methods, see Rima D. Apple, *Mothers and Medicine: A Social History of Infant Feeding, 1890–1950* (Madison: University of Wisconsin Press, 1987), 24–29; Harvey Levenstein, "'Best for Babies' or 'Preventable Infanticide'? The Controversy over Artificial Feeding of Infants in America, 1880–1920," *Journal of American History* 70 (1983): 81–84; Richard A. Meckel, *Save the Babies: American Public Health Reform and the Prevention of Infant Mortality, 1850–1929* (Baltimore: Johns Hopkins University Press, 1990), 57–59; and Edwards A. Park and Howard A. Mason, "Luther Emmett Holt," in Veeder, *Pediatric Profiles*, 44–48.

5. Levenstein, 76.

6. Park and Mason, 45.

7. Rotch, *Pediatrics*, 288.

8. Rotch, *Pediatrics*, 297–99, 308–12.

9. Thomas M. Rotch, "General Principles Underlying All Good Methods of Infant Feeding," *Boston Medical and Surgical Journal* 129 (1893): 505.

10. On oxygen therapy, see "History of Oxygen Therapy and Retrolental Fibroplasia," *Pediatrics*, vol. 57, supp. 4 (1976): 593–594.

11. Little research has been done regarding the history of oxygen therapy. In spite of the early interest expressed by such figures as Joseph Priestly and Humphrey Davies in Britain, the cheap manufacture and provision of oxygen awaited the commercialization of the Tessie du Môtay process after 1870. In New York, the newly organized Oxygen Gas Company was so successful that by 1872, the second year of its operation, it had provided over four hundred thousand gallons of oxygen to physicians and hospitals. Most commonly, physicians used the gas as a stimulant or tonic for chronic disease, as well as a palliative applied in short bursts for respiratory diseases such as pneumonia. Awareness of the toxicity of oxygen originated with Priestly, whose observations of the effect of immersing a smoldering body into a tank of oxygen provoked concern that the gas might excite excessive action and literally cause the patient to live too fast. See Andrew H. Smith, "Oxygen Gas as a Remedy in Disease," *New York Medical Journal* 11 (1870): 113–68; and J. Henry Davenport, "Oxygen as a Remedial Agent," *Boston Medical and Surgical Journal* 87 (1872): 61–64.

12.His case records of one premature infant in an incubator report that Rotch typically applied oxygen for five-to-ten-minute bursts, two to three times daily, early in the infant's life. His schedule suggests that he initially provided the gas on a regular regimen, which he gradually weaned down, following which the gas was reserved for treating symptomatic episodes of cyanosis. The records are reproduced in Rotch, *Pediatrics*, 306–7.

92. BH-NY, *AR* (1904), 15; the incorporation of the Lion incubator is cited in a later review; BH-NY, *AR* (1912), 21.

93. BH-NY, *AR* (1905), 14.

94. BH-NY, *AR* (1908), 25.

95. BH-NY, *AR* (1909), 20–22.

96. Isaac A. Abt, "Individual Prophylaxis in Children's Hospitals," *Journal of the American Medical Association* 59 (1912): 1687. The metaphor of the operating room was indeed appropriate, for asepsis and the contact theory of infection were increasingly to guide further reform. Contact theory made great strides during the early 1900s with the elucidation of carrier states for infections such as typhoid. The publication of Charles V. Chapin's *The Sources and Modes of Infection* (New York: John Wiley and Sons, 1910) marked an important milestone in its dissemination; see C. E. A. Winslow, *The Conquest of Epidemic Disease* (Princeton: Princeton University Press, 1943), 337–46, 370–73.

97. The drive toward partitioning the wards originated in the "box" system developed at the Pasteur Institute in Paris, designed to separate patients into self-enclosed alcoves containing separate utensils and equipment. Some partitions were open at the top, reflecting the contact, rather than atmospheric, theory of infection. The system soon spread to German hospitals in the early 1900s; incubator rooms emerged simultaneously. For a review by a partisan of the contact theory, see Henry Koplik, "Hospitals for the Care of Infants and Children, and the Methods of Prevention of Infection," *Archives of Pediatrics* 28 (1911): 728–39; 29 (1912): 5–11. In the United States, the wave of infant hospital reform that Holt foreshadowed spread across the country between 1910 and 1920. No overview is available, but many articles appeared describing specific hospitals. Examples include Charles A. Coolidge (on Boston Children's Hospital), "The Architecture of the Children's Hospital," *Boston Medical and Surgical Journal* 170 (1914): 481–83; H. J. Gerstenberger, Abram Garfield, and S. S. Goldwater, "Babies' and Children's Hospital of Cleveland," *Modern Hospital* 17 (1921): 187–94; A. L. Goodman, "A Modern Children's Pavilion," *Archives of Pediatrics* 32 (1915): 684–92; and Borden S. Veeder, "The St. Louis Children's Hospital," *Modern Hospital* 5 (1915): 387–94.

99. BH-NY, *AR* (1910), 18–19.

99. L. Emmett Holt, *Diseases of Infancy and Childhood,* 6th ed. (New York: D. Appleton, 1911), 12–13; BH-NY, *AR* (1911), 42.

100. BH-NY, *AR* (1910), 17–18.

101. BH-NY, *AR* (1912), 22.

102. L. Emmett Holt, "The Children's Hospital, the Medical School, and the Public," *Johns Hopkins Hospital Bulletin* 24 (1913): 92.

103. Julius H. Hess, *Premature and Congenitally Diseased Infants* (Philadelphia: Lea and Febiger, 1922), 226–28.

104. LaFetra, 26–27.

105. Walter Lester Carr, "A Clinical Report of Simple Methods in the Care of Premature Babies," *Archives of Pediatrics* 38 (1921): 401–4.

106. Kenneth D. Blackfan and Constantin Yaglou, "The Premature Infant: A Study of Atmospheric Conditions on Growth and Development," *American Journal of Diseases of Children* 46 (1933): 1175–1236.

107. Pierre Budin, *The Nursling: The Feeding and Hygiene of Premature and Full-Term Infants,* trans. William J. Maloney (London: Caxton Publishing, 1907), 13.
108. BH-NY, *AR* (1890), 6.
109. BH-NY, *AR* (1913), 15.

8. The Eclipse of the Incubator

1. On the European campaigns, see Alisa Klaus, *Every Child a Lion: The Origins of Maternal and Infant Health Policy in the United States and France, 1890–1920* (Ithaca: Cornell University Press, 1993); Deborah Dwork, *War is Good for Babies and Other Children: A History of the Infant and Child Welfare Movement in England, 1898–1918* (London: Tavistock, 1987).
2. Susan Tifflin, *In Whose Best Interest? Child Welfare Reform in the Progressive Era* (Westport, Conn.: Greenwood Press, 1982). A useful introduction to social reform in the Progressive Era is Arthur S. Link and Richard L. McCormick, *Progressivism* (Arlington Heights, Ill.: Harlan Davidson, 1983), especially 79–80, with regard to children.
3. Richard A. Meckel, *Save the Babies: American Public Health Reform and the Prevention of Infant Mortality, 1850–1929* (Baltimore: Johns Hopkins University Press, 1990), 63–91, 107–10.
4. Anne Firor Scott, *Natural Allies: Women's Associations in American History* (Urbana: University of Illinois Press, 1993), especially 141–74.
5. Sarah J. McNutt, "Medical Women, Yesterday and Today," *Medical Record* 94 (1918): 138–39.
6. Regina Markell Morantz-Sanchez, *Sympathy and Science: Women Physicians in American Medicine* (New York: Oxford University Press, 1985), 266–311; Susan M. Reverby, *Ordered to Care: The Dilemma of American Nursing, 1850–1945* (Cambridge: Cambridge University Press, 1987), 109–10. Sarah McNutt, the founder of the New York Babies' Hospital, recalled that her mentor, Elizabeth Blackwell, encouraged her to pursue pediatrics; since men had neglected the field, she would not have to worry about whether others approved. McNutt, 138.
7. For an overview of the settlement movement, see Allen Davis, *Spearheads for Reform: The Social Settlements and the Progressive Movement, 1890–1914* (New York: Oxford University Press, 1967).
8. No biographies yet exist for either of these two important figures. Each, however, left memoirs. See Lillian D. Wald, *The House on Henry Street* (New York: Henry Holt, 1915), especially 52–53, on infant care; and S. Josephine Baker, *Fighting for Life* (New York: Macmillan, 1939). On Baker's visiting nurse program, see Leona Baumgartner, "Sara Josephine Baker," in *Notable American Women,* ed. E. T. James, J. W. James, and P. S. Boyer (Cambridge: Belknap Press, 1971), 85–86; and Meckel, 134–39.
9. Two settlement alumni, Lillian Wald and Florence Kelly, proposed the idea of the Children's Bureau and its first two directors, Julia Lathrop and Grace Abbot, were both graduates of Hull House. Nancy P. Weiss, "Save the Children: A History of the Children's Bureau, 1903–1918," (Ph.D. diss., University of California, Los Angeles, 1974).
10. Wilbur C. Phillips, "A Plan for Reducing Infant Mortality in New York City," *Medical Record* 73 (1908): 891.

11. Meckel, 93–99.

12. George Newman, *Infant Mortality: A Social Problem* (London: Methuen, 1906).

13. Baumgartner, 85; Meckel, 134–40.

14. Jennings C. Litzenberg, "The Care of Premature Infants, with Special Reference to the Use of Home-Made Incubators," *Journal of the Minnesota Medical Association* 28 (1908): 87.

15. The literature on eugenics is extensive; one of the most helpful recent works is Daniel Kevles, *In the Name of Eugenics: Genetics and the Uses of Human Heredity* (New York: Alfred A. Knopf, 1985). On the implications of eugenics for infant mortality reformers, see Meckel, 116–18.

16. The AASPIM included a section on eugenics from the outset, and many of its most distinguished members expressed some degree of sympathy with eugenic ideals. See Abraham Jacobi, "Address," *Transactions of the American Association for the Study and Prevention of Infant Mortality* 1 (1910): 43. William Welch, the dean of Johns Hopkins, expressly remarked that the organization crusade would do best to avoid causes of infant mortality "which to some seem unavoidable," particularly premature birth; Welch, "Address," *Transactions of the American Association for the Study and Prevention of Infant Mortality* 1 (1910): 52.

17. Newman, 79–81.

18. Ibid., 84.

19. U.S. Dept. of Commerce and Labor, Bureau of the Census, *Mortality Statistics, 1910* (Washington, D.C.: Government Printing Office, 1913), 533.

20. Meckel, 166–71.

21. The studies of Adolphe Pinard in France had confirmed an adverse association between maternal employment and birth weight as early as the 1890s, although the causal relationship between the two remained uncertain. In the Anglo-American world, the British obstetrician John W. Ballantyne became a particularly influential advocate of prenatal care and of antepartum hospitalization to assure proper rest and care. J. W. Ballantyne, "A Plea for a Pro-Maternity Hospital," *British Medical Journal* 1 (1901): 813–14; J. W. Ballantyne, "Hygiene of the Mother before the Birth of Her Child," *Practitioner* 75 (1905): 433–41.

22. John M. Glenn, Lillian Brandt, and F. Emerson Andrews, *Russell Sage Foundation, 1907–1946* (New York: Russell Sage Foundation, 1947), 1:106; Meckel, 164–65.

23. U.S. Dept. of Labor, Children's Bureau, *Baby-Saving Campaigns: A Preliminary Report on What American Cities Are Doing to Prevent Infant Mortality,* Bureau Pub. 3 (Washington, D.C.: Government Printing Office, 1913), 38–39.

24. Mrs. William Lowell Putnam, "The Most Efficient Means of Preventing Infant Mortality," *American Journal of Obstetrics and Diseases of Women and Children* 78 (1918): 104.

25. An overview of early prenatal care was provided by Mrs. Max [Mary Mills] West, "The Development of Prenatal Care in the United States," *Transactions of the American Association for the Study and Prevention of Infant Mortality* 5 (1914): 69–108.

26. U.S. Dept. of Labor, Children's Bureau, *Prenatal Care,* Bureau Pub. 4, Mrs. Max West (Washington, D.C.: Government Printing Office, 1913), 8–14; U.S. Dept. of

Labor, Children's Bureau, *Infant Care,* Bureau Pub. No. 8, Mrs. Max West (Washington, D.C.: Government Printing Office, 1914), 57.

27. Mrs. Max West, "The Prenatal Problem and the Influences Which May Favorably Affect This Period of the Child's Growth," *Transactions of the American Association for the Study and Prevention of Infant Mortality* 6 (1915): 219.

28. Henry J. Hibbs, *Infant Mortality: Its Relation to Social and Industrial Conditions* (Concord, N.H.: Rumford Press, 1916), 22; Putnam, 103.

29. West, "Prenatal Problem," 224.

30. Lillian Wald was one of the earlier observers to see the problem of infant hospitalism in terms of a maternal rather than environmental, explanation; see Wald, "The District Nurses' Contribution to the Reduction of Infant Mortality," *Bulletin of the American Academy of Medicine* 11(1909): 376.

31. Henry D. Chapin, "The Proper Management of Foundlings and Neglected Infants," *Medical Record* 79 (1911): 284.

32. Henry D. Chapin, "Are Institutions for Infants Necessary?" *Transactions of the American Association for the Study and Prevention of Infant Mortality* 5 (1914): 127.

33. S. Josephine Baker, "The Possible Reduction of Infant Mortality among Sub-Normal Institutional Babies," *Transactions of the American Association for the Study and Prevention of Infant Mortality* 6 (1915): 247.

34. S. Josephine Baker, *Fighting for Life,* 120.

35. S. Josephine Baker, "Sub-Normal Babies," 248–49.

36. Ibid., 251.

37. Ibid. Interestingly, in a later recollection of how the incident had made her a firm believer in "old-fashioned sentimental mothering," Baker explicitly employed the metaphor of technology in condemning the hospital. She wrote that the experience had led her to conclude that the real problem was "the foundling hospital's nurse who turned the foundling over at the right time and gave him the best of care with all the impersonal efficiency of a well-intentioned machine"; see Baker, *Fighting for Life,* 121. Chapin employed similar imagery with respect to elementary education, noting that "the child soon becomes a little machine, to be wound up at each hour of the day for some special labor or service"; see Henry Dwight Chapin, *Vital Questions* (New York: Thomas Y. Crowell, 1905), 105.

38. Abraham Flexner, "Biographical Memoir," in *A. F. Hess: Collected Writings,* (Springfield, Mass.: Charles C. Thomas, 1936), 1:xv.

39. A. F. Hess, "Institutions as Foster Mothers for Infants," *Transactions of the American Association for the Study and Prevention of Infant Mortality* 6 (1915): 262.

40. S. Josephine Baker, discussion following West, "Development of Prenatal Care," 112.

41. "Feminist," in this case, is being used in a very broad sense, to encompass the many women of the nineteenth century who sought to increase the scope of their social roles outside the home while still maintaining an outlook distinct from that of men. In this sense, women physicians could argue that they possessed natural gifts for compassion and nurturing that ideally suited them to be healers. This was particularly true when the patients were children. See Morantz-Sanchez, 4–5.

42. Budin, who was more sympathetic than many of his peers, could still be highly

paternalistic, especially in his earlier years. In a 1891 manual for midwives, he asserted that "the primary role of woman is a role of reproduction." Pierre Budin and E. Crouzat, introduction to *La pratique des accouchements a l'usage des sages-femmes* (Paris: Octave Doin, 1891).

43. Meckel, 100–101.

44. Meckel, 178–99. On the Progressive campaign for government-insured sickness as well as maternity insurance, see Ronald L. Numbers, *Almost Persuaded: American Physicians and Compulsory Health Insurance, 1912–1920* (Baltimore: Johns Hopkins University Press, 1978).

45. The targeted communities included Johnstown, Pa. (1915), Manchester, N.H. (1917), Waterbury, Conn. (1918), Brockton, Mass. (1918), Saginaw, Mich. (1919), Akron, Ohio (1920), New Bedford, Mass. (1920), Baltimore, Md. (1921), and Gary, Ind. (1923). The studies were later summarized in Robert Morse Woodbury, *Infant Mortality and Its Causes* (Baltimore: William and Wilkins, 1926).

46. Woodbury, 145–46.

47. Children's Bureau members solicited and personally answered letters from women throughout the country. Many came from rural areas with little access to medical advice. For an overview, see Molly Ladd-Taylor, *Raising a Baby the Government Way: Mothers' Letters to the Children's Bureau, 1915–1932* (New Brunswick, N.J.: Rutgers University Press, 1986).

48. Mrs. George Sass to Julia Lathrop, Racine, Wisc., received 10 Mar. 1916, U.S. Dept of Labor, Children's Bureau, Correspondence 4-2-1-4, National Archives, Washington, D.C.

49. Mrs. Helen Bentcliff to the Children's Bureau, Chicago, Ill., 28 Feb. 1916, U.S. Dept. of Labor, Children's Bureau, Correspondence 4-2-0-3.

50. Julia Lathrop, to Alice Hamilton, Washington, D.C., 2 Mar. 1916, U.S. Dept. of Labor, Children's Bureau, Correspondence 4-2-0-3.

51. Meckel, 200–219.

52. A Children's Bureau report in 1931 noted that recent studies had suggested that the prevention of neonatal morbidity and mortality was far from simple and suggested that energy would be better shifted toward improving treatment. Since knowledge of care far exceeded standards in actual practice, it concluded that "to make such methods available would appear to be the first and most obvious way to bring about a reduction in neonatal mortality." U.S. Dept. of Labor, Children's Bureau, *Nineteenth Annual Report of the Chief of the Children's Bureau to the Secretary of Labor, 1931* (Washington, D.C.: Government Printing Office, 1931), 3. For another assessment of why attention shifted from prevention to treatment around 1930, see Clifford G. Grulee, "The Effect of Prenatal Care upon the Infant," *Transactions of the American Child Hygiene Association* 6 (1929): 90–91.

53. J. Morris Slemans, *John Whitridge Williams: Academic Aspects and Bibliography* (Baltimore: Johns Hopkins University Press, 1935), 7, 42–43, 49; D. N. Danforth, "Contemporary Titans: Joseph Bolivar DeLee and John Whitridge Williams," *American Journal of Obstetrics and Gynecology* 120 (1974): 577–88.

54. Slemans, 23–25; Judith Walzer Leavitt, *Brought to Bed: Childbearing in America, 1750–1950* (New York: Oxford University Press, 1986), 179–80, 183–86.

55. J. Whitridge Williams, "Significance of Syphilis in Prenatal Care and in the Causation of Foetal Death," *Bulletin of the Johns Hopkins Hospital* 31 (1920): 141.

56. J. Whitridge Williams, "What the Obstetrician Can do to Prevent Infant Mortality," *Transactions of the American Association for the Study and Prevention of Infant Mortality* 1 (1910): 190.

57. Allan M. Brandt, *No Magic Bullet: A Social History of Venereal Disease in the United States since 1880* (New York: Oxford University Press, 1987), 40.

58. J. Whitridge Williams, "The Limitations and Possibilities of Prenatal Care, Based upon the Study of 705 Foetal Deaths Occurring in 10,000 Admissions to the Obstetrical Department of the Johns Hopkins Hospital," *Transactions of the American Association for the Study and Prevention of Infant Mortality* 5 (1914): 33–34; statistics are from table, 34. The address was reprinted in *Journal of the American Medical Association* 64 (1915): 95–101.

59. Williams, "Limitations of Prenatal Care," 43; Williams, "Significance of Syphilis," 142.

60. Williams, "Limitations of Prenatal Care," 37, 42; John F. Moran, "The Endowment of Childhood from the Obstetric Standpoint," *Journal of the American Medical Association* 65 (1915): 227. On the struggle between obstetricians and midwives see Williams' influential paper "The Midwife Problem and Medical Education in the United States," *Transactions of the American Association for the Study and Prevention of Infant Mortality* 2 (1911): 165–94, and Judy Barrett Litoff, *American Midwives: 1860 to the Present* (Westport, Conn.: Greenwood Press, 1978).

61. Brandt, 46, 161.

62. Williams, "Significance of Syphilis," 142, 145.

63. Williams, "The Value of the Wasserman Reaction in Obstetrics, Based upon the Study of 4,547 Consecutive Cases," *Bulletin of the Johns Hopkins Hospital* 31 (1920): 336–37.

64. Williams, "Significance of Syphilis," 145.

65. L. Emmett Holt and Ellen C. Babbitt, "Institutional Mortality of the Newborn: A Report on Ten Thousand Consecutive Births at the Sloane Hospital for Women, New York," *Transactions of the American Association for the Study and Prevention of Infant Mortality* 5 (1914): 151–52.

66. Ibid., 154.

67. Ibid., 156.

68. Ibid., 154.

69. Jacobi, "Address," 43.

70. Holt and Babbitt, 158.

71. Stillbirth continues to occur in 25–50 percent of syphilitic pregnancies. The rate of perinatal death (occurring soon after birth) in congenital syphilis is more controversial. Although perinatal death rates on the order of 25–30 percent have often been cited, these figures come from older studies and may largely reflect the mortality of prematurity alone. See *Infectious Diseases of Children*, 9th ed., ed. Saul Krugman, Samuel L. Katz, Anne A. Gershon, and Catherine Wilfert (St. Louis: C. V. Mosby, 1992), 430.

72. Charles E. Rosenberg, *The Care of Strangers: The Rise of America's Hospital System* (New York: Basic Books, 1987), 404 n. 29.

73. Grace Abbott, "Administration of the Sheppard-Towner Act for Maternal Care," *Transactions of the American Child Hygiene Association* 13 (1922): 200.

74. The best study of the transformation of American childbirth from the perspective of women is Judith Walzer Leavitt, *Brought to Bed*. Also see Richard Wertz and Dorothy Wertz, *Lying-in: A History of Childbirth in America*, expanded ed. (New Haven: Yale University Press, 1989); and Litoff.

75. L. Emmett Holt, "The Children's Hospital, the Medical School, and the Public," *Johns Hopkins Hospital Bulletin* 24 (1913): 89.

76. Abraham Jacobi, "The Best Means of Combating Infant Mortality," *Journal of the American Medical Association* 58 (1912): 1737.

77. Jennings C. Litzenberg, "How the Pediatrician and Obstetrician Can Cooperate," *American Journal of Obstetrics and Diseases of Women and Children* 77 (1918): 465–66. On the AATDC and its campaign to win pediatric control of the obstetric nurseries, see Sydney A. Halpern, *American Pediatrics: The Social Dynamics of Professionalism, 1880–1980* (Berkeley: University of California Press, 1988), 70–71.

78. One of the few American obstetricians to defend their control of the nursery in print was A. N Creadick, of New Haven, who appealed to breast milk in justifying his case. Creadick, "Conduct of the Nursery in an Obstetrical Clinic," *Transactions of the American Child Hygiene Association* 11 (1920): 31, 40–41.

79. Ballantyne used the term to criticize neglect of the newborn and believed that the obstetrician was best qualified to supervise the infant's care prior to weaning. J. W. Ballantyne, "Where Obstetrics and Paediatrics Meet: Infant Welfare," *International Clinics*, 26th ser., 4 (1916): 96.

80. Litzenberg, "How the Pediatrician and Obstetrician Can Cooperate," 463; Barnet E. Bonar, "Problems of the Newborn," *Journal of Pediatrics* 1 (1932): 92.

81. Howland, himself a resident under Holt, led the first completely full-time clinical department at Johns Hopkins; many of his residents later went on to head pediatric departments of their own. See A. McGehee Harvey, Gert H. Brieger, Susan L. Abrams, and Victor A. McKusick, *A Model of Its Kind* (Baltimore: Johns Hopkins University Press, 1989), 55–57.

82. Williams, "What the Obstetrician Can Do," 200.

83. Overall, forty-six hundred of the ten thousand mothers in Williams's series were African-American. The mortality rates for black and white newborns were 9.4 percent and 5.1 percent respectively; the corresponding rates of syphilis were 35 percent and 14 percent. Williams, "Limitations of Prenatal Care," 33, 35.

84. Williams, "Limitations of Prenatal Care," 33, 38.

85. Helen Ford, "Racial Factors in Relation to Neonatal Mortality," *Nation's Health* 6 (1924): 254–55; John Hall Mason Knox, "Morbidity and Mortality in the Negro Infant," *Transactions of the American Pediatric Society* 36 (1924): 46–51.

86. C. F. Wilinsky, "A Study of Maternal and Infant Mortality in Boston," *Transactions of the American Child Hygiene Association* 6 (1929): 133.

87. Neonatologist William Silverman has written that the practice only stopped with the creation of the Apgar score in 1952 for assessing asphyxia at one and five minutes of life. Marginal newborns no longer could be labeled as stillborn. William A. Silverman, "Overtreatment of Neonates? A Personal Retrospective," *Pediatrics* 90 (1992): 973.

88. Litzenberg, "Care of Premature Infants," 87–91; Litzenberg, "How the Pediatrician and Obstetrician Can Cooperate," 463–68; Jennings C. Litzenberg, "Better Newly-born Pediatrics," *Southern Medical Journal* 17 (1924): 378–81.

89. J. P. Sedgwick, quoted in discussion, "Conduct of the Nursery in an Obstetrical Clinic," *Transactions of the American Child Hygiene Association* 11 (1920): 44.

90. Litzenberg, "How the Pediatrician and Obstetrician Can Cooperate," 466.

91. Published information regarding the geographical distribution of incubator rooms is limited. In his 1922 textbook, Julius Hess cited three institutions as having exemplary incubator rooms: Washington University in St. Louis, the University of California at San Francisco, and his own institution, Michael Reese of Chicago. See Julius H. Hess, *Premature and Congenitally Diseased Infants* (Philadelphia: Lea and Febiger, 1922), 226–28; also (for San Francisco) Paul Cook, "A Clinical Study of the Premature Infant," *Archives of Pediatrics* 38 (1921): 201–16. The only other incubator rooms, besides Holt's, to appear in the published literature before the late 1920s were at New York's Bellevue Hospital (Linnaeus E. LaFetra, "The Hospital Care of Premature Infants," *Archives of Pediatrics* 34 (1917): 26) and the Manhattan Maternity Hospital (Walter Lester Carr, "A Clinical Report of Simple Methods in the Care of Premature Babies," *Archives of Pediatrics* 38 (1921): 401). Both incubator rooms, especially that at Bellevue, appear to have been far cruder than the rooms in St. Louis, Chicago, and San Francisco.

92. For Hess's biography, see Alwin C. Rambar, "Julius Hess, M.D.," in *Historical Review and Recent Advances in Neonatal and Perinatal Medicine*, ed. G. F. Smith, P. N. Smith, and D. Vidyasagar (Chicago: Mead Johnson Nutritional Division, 1983), 2:161–64. Also see obituaries in *American Journal of Diseases of Children* 91 (1956): 289; and *Journal of the American Medical Association* 159 (1955): 1227.

93. Julius H. Hess, "A Study of the Caloric Needs of Premature Infants," *American Journal of the Diseases of Children* 2 (1911): 302–14.

94. Obituary, *American Journal of Diseases of Children* 91 (1956): 289; Silverman, "Incubator-Baby Side Shows," 136–37; L. Joseph Butterfield, "The Incubator Doctor in Denver: A Medical Missing Link," in *The 1970 Denver Westerner Brand Book* (Denver: Westerners, 1971), 358–59.

95. Isaac A. Abt, *Baby Doctor* (New York: Whittlesley House, 1944), 105; *All Our Lives: A Centennial History of Michael Reese Hospital and Medical Center, 1881–1981*, ed. Sarah Gordon (Chicago: Michael Reese, 1981), 86.

96. Translated newspaper clipping from Chicago Jewish newspaper *Westen und Daheim, 16* June 1907, in Gordon, 74.

97. Gordon, 9.

98. Though the role of anti-Semitism in the incubator story is hard to document, its significance in American medicine deserves further exploration. Certainly, the Chicago Jewish communities were essential to supporting the work of DeLee and Hess. Far less clear is whether eugenic concerns made other prospective donors skeptical of their efforts to save premature and weak infants, most of whom came from lower-income and Eastern European backgrounds. I am grateful to the reviewer of this manuscript for suggesting this issue.

99. Julius H. Hess, "An Electric-Heated Water-Jacketed Infant Incubator and Bed,

for Use in the Case of Premature and Poorly Nourished Infants," *Journal of the American Medical Association* 64 (1915): 1068–69.

100. Julius H. Hess, "The Care of Premature Infants," *Medical Clinics of North America* 3 (May 1920): 1722.

101. Hess, *Premature and Congenitally Diseased Infants*, 228.

102. Hess, "Care of Premature Infants," 1717, 1721.

103. Julius H. Hess, "Heated Bed for Transportation of Premature Infants," *Journal of the American Medical Association* 80 (1923): 1313. Hess explicitly acknowledged DeLee in the first account of his transport incubator; see Hess, *Premature and Congenitally Diseased Infants*, 230–31.

104. Natale P. Solway, "The Story of the Infants' Aid Society," in *Care of the Premature Infant*, ed. Evelyn C. Lundeen and Ralph H. Kundstadter, (Philadelphia: J. B. Lippincott, 1958), 306–9; Julius H. Hess, "Speech to Infants' Aid" (date unspecified but close to 1950), Julius H. Hess Papers, Regenstein Library, University of Chicago, Chicago.

105. Hess, "Care of Premature Infants," 1709.

106. Gordon, 93; Solway, 307.

107. For descriptions, see Gordon, 88; Rambar, 163; and Julius H. Hess, "History of the Organization and Development of the Premature Infant Station," in *The Physical and Mental Growth of Prematurely Born Children*, ed. Julius H. Hess, George J. Mohr, and Phyllis Bartelme (Chicago: University of Chicago Press, 1934), 3–11; on wet nurses see Hess, *Premature and Congenitally Diseased Infants*, 118–24.

108. Hess, 103.

109. Hess, *Premature and Congenitally Diseased Infants*, 381–87; Julius H. Hess and I. M. Chamberlain, "Premature Infants: A Report of Two Hundred and Sixty-six Consecutive Cases," *American Journal of Diseases of Children* 34 (1927): 571–84; Hess, Mohr, and Bartelme.

110. Lundeen and Kundstadter, v.

111. Hess, "Care of Premature Infants," 1721–22; Hess, *Premature and Congenitally Diseased Infants*, 146.

112. For Lundeen's biography, see L. Joseph Butterfield, "Historical Perspectives of Neonatal Transport," *Pediatric Clinics of North America* 40 (Apr. 1993): 231–32.

113. Rambar, 163. For a similar portrayal, see Joseph Calvin, quoted in Gordon, 93.

114. Julius H. Hess and Evelyn Lundeen, *The Premature Infant: Its Medical and Nursing Care* (Philadelphia: J. B. Lippincott, 1941), 27.

115. Julius H. Hess, "Oxygen Unit for Premature and Very Young Infants," *American Journal of the Diseases of Children* 47 (1934): 916–18; William M. Boothby, "Miniature Oxygen Chamber for Infants: A Modification of the Hess Incubator," *Proceedings of the Staff Meetings of the Mayo Clinic* 9 (1934): 129–31. For a criticism of the device, see Scanlan-Morris Company, Report on Tests of Hess Infant Incubator and Bed with Hess Infant Oxygen Unit," 12 Apr. 1940, Hess Papers. The most successful of the new generation of closed, high-oxygen incubators was the Isolette or Chapple Incubator; Charles C. Chapple, "Controlling the External Environment of Premature Infants in an Incubator," *American Journal of Diseases of Children* 56 (1938): 459–60; Thomas E. Cone, *History of the Care and Feeding of the Premature Infant* (Boston: Lit-

tle, Brown, 1985), 70–72. The Hess beds, incidentally, escaped the retrolental fibroplasia epidemic, presumably because they were not airtight; Rambar, 164. For an excellent overview of oxygen and the premature infant, see William A. Silverman, *Retrolental Fibroplasia: A Modern Parable* (New York: Grune and Stratton, 1980), 43–51.

ESSAY ON SOURCES

Physicians and historians interested in the origins of neonatology will likely find a medical center library endowed with a substantial collection of older journals more useful than a trip to an archive. Nevertheless, several sets of manuscripts are worth noting. Julius Hess's papers at the Joseph Regenstein Library of the University of Chicago are particularly extensive, though most address his career after 1930. They include numerous personal letters, drafts for his books, and a number of items related to Martin Couney. The marvelous collection of Joseph DeLee's papers held at Northwestern Memorial Hospital Archives in Chicago provides great insight into DeLee's obstetric career including his incubator station. Thomas Rotch is represented by a small collection at the Francis Countway Library at Harvard Medical School; to the best of my knowledge, L. Emmett Holt, Henry Chapin, John Zahorsky, and Martin Couney have left no collected papers currently available in public collections. Documentation from parents is in even shorter supply. A fascinating exception is the extensive collection of mothers' letters from 1914 to 1932 preserved in the U.S. Department of the Interior Children's Bureau Correspondence at the National Archives.

On the French side, the archives of Assistance Publique in Paris contain birth and death registers for the Maternité's *service des débiles* from 1895 to 1903. Though interesting from a demographic standpoint, these documents offer little medical information not reproduced in published articles by Pierre Budin, Adolphe Pinard, and Charles Porak cited in the text footnotes. Assistance Publique, however, also holds numerous obscure reprints on Budin, Pinard, and Stéphane Tarnier, as well as reports from the Commission on Depopulation relevant to premature infant mortality. The most extensive collection of papers I could locate pertaining to Budin is privately owned by Dr. Paul Toubas of Oklahoma City. These have limited bearing on premature infant care but contain many letters and documents that will be of great interest for anyone investigating late-nineteenth-century French obstetrics.

In lieu of personal papers, hospital annual reports can provide considerable insight into the actual practice of early hospital pediatrics. Some, such as those of the New York Babies' Hospital and New York Postgraduate Hospital (both available in the National Library of Medicine in Bethesda, Maryland), devote considerable attention to incubator technology and premature infant therapy. The annual reports of DeLee's Chicago Lying-in Hospital, available in his papers as well as through the Chicago Historical Society, offer an especially illuminating account of his incubator station. In contrast, annual reports of the Boston Lying-in Hospital, the New York Infant Asylum, the New York Foundling Hospital, the Nursery and Child's Hospital (New York), and the Lying-in Hospital of New York yield little information on the care of premature infants in the late 1800s or early 1900s.

Fortunately for the historian, medical textbooks and journal articles at the turn of the century provided more personal testimony than do their counterparts today, often describing the authors' particular experiences with premature infants rather than merely recounting general standards of care. For example, Pierre Budin's *The Nursling: The Feeding and Hygiene of Premature and Full-Term Infants,* translated by William J. Maloney (London: Caxton Publishing, 1907), foremost among the French textbooks on newborn medicine, consists of lectures with considerable historical content recounting the author's actual experiences with premature infant care. A similar point could be made for several of the other key publications related to early incubator technology: Alfred Auvard's "De la couveuse pour enfants," *Archives de tocologie des maladies des femmes et des enfants nouveau-nés* 10 (1883): 577–609; Paul Berthod's *Les enfants nés avant terme: La couveuse et la gavage à la Maternité de Paris* (Paris: G. Rougier, 1887); John Zahorsky's *Baby Incubators: A Clinical Study of the Premature Infant* (St. Louis: Courier of Medicine Press, 1904); and Julius H. Hess's *Premature and Congenitally Diseased Infants* (Philadelphia: Lea and Febiger, 1922). The two leading American pediatric textbooks of the early twentieth century, L. Emmett Holt's *The Diseases of Infancy and Childhood* (New York: D. Appleton, 1897) and Thomas M. Rotch's *Pediatrics: The Hygienic and Medical Treatment of Children* (Philadelphia: J. B. Lippincott, 1895), taken with their subsequent editions, deserve close scrutiny as well.

Medical journal articles constitute the heart of this study. Yet their access at the turn of the century is somewhat problematic. The *Index-Catalogue of the Library of the Surgeon General's Office, U.S. Army* (Washington, D.C.: Government Printing Office, 1880–1932; New York: Johnson Reprint Corp., 1972) is spotty in covering early-twentieth-century premature infant articles falling between its second and third series. *Index Medicus* (New York: F. Leypoldt, 1879–1927; New York: Johnson Reprint Corp., 1967) likewise skips the years 1900–2 between its first two series (1879–99 and 1903–20), a deficiency for which the *Bibliographia Medica* (Paris: Institut de Bibliographie, 1900–2) only partially compensates. I therefore supplemented these standard bibliographical sources by searching the annual indices of some of the most important pediatric and medical journals of the time. These included: *American Journal of Nursing* (1901–22), *American Journal of Obstetrics and Diseases of Women and Children* (1879–1919), *Archives de médecine des enfants* (1898–1910), *Archives of Pediatrics* (1884–1922), *Bulletin de l'Académie de Médecine*

(1872–1910), *Journal of the American Medical Association* (1883–1928), *La presse médicale* (1896–1908), *Pediatrics* (1896–1917), *Transactions of the American Association of Obstetricians and Gynecologists* (1888–1919), *Transactions of the American Association for the Study and Prevention of Infant Mortality* (1910–18), *Transactions of the American Child Hygiene Association* (1919–22), *Transactions of the American Child Health Association* (1923–29), *Transactions of the American Gynecological Society* (1877–1901), and *Transactions of the American Pediatric Society* (1889–1924). Cross-references led to many further sources.

Most of the available secondary literature on the history of the premature infant nursery and newborn intensive care has been written by clinicians. The starting point for anyone interested in exploring the subject should be Thomas E. Cone's *History of the Care and Feeding of the Premature Infant* (Boston: Little, Brown, 1985), which is especially notable for its attention to technical and scientific advances and its superb bibliography. Whereas Cone's concern lies in documenting the progress of newborn care and technology, William A. Silverman provides a more critical perspective in his excellent *Retrolental Fibroplasia: A Modern Parable* (New York: Grune and Stratton, 1980). This book explores the consequences of high-oxygen therapy in the 1940s and 1950s and provides the best introduction to neonatal care between Julius Hess's day and the rise of newborn intensive care after 1960. Silverman has also written other insightful articles elucidating the lessons of history for neonatology today, including "Neonatal Pediatrics at the Century Mark," *Perspectives in Biology and Medicine* 32 (1989): 159–70, and "Overtreatment of Neonates? A Personal Retrospective," *Pediatrics* 90 (1992): 971–76. Also worthy of note is L. Joseph Butterfield's "Historical Perspectives of Neonatal Transport," *Pediatric Clinics of North America* 40 (Apr. 1993): 221–39.

On the use of premature infants in fairs and expositions, see William A. Silverman's "Incubator-Baby Side Shows," *Pediatrics* 64 (1979): 127–41, and Butterfield's "The Incubator Doctor in Denver: A Medical Missing Link," in *The 1970 Denver Westerners Brand Book* (Denver: Westerners, 1971). Less useful are the essays in *Historical Review and Recent Advances in Neonatal and Perinatal Medicine,* 2 vols., ed. G. F. Smith, P. N. Smith, and D. Vidyasagar (Chicago: Mead Johnson, 1983), a book whose general focus, in spite of its title, is on the present.

The history of American pediatrics and child health, once a neglected stepchild of medical history, has come into its own in the past ten years. The foremost work to date is Richard A. Meckel's *Save the Babies: American Public Health Reform and the Prevention of Infant Mortality, 1850–1929* (Baltimore: Johns Hopkins University Press, 1990), a sophisticated yet highly readable analysis of the late-nineteenth and early-twentieth century infant mortality campaign that incorporates much of the early history of pediatrics. *Fatal*

Years: Child Mortality in Late-Nineteenth-Century America (Princeton: Princeton University Press, 1991), by Samuel Preston and Michael Haines, provides a demographic analysis that assigns primary importance to medical science in the reduction of infant mortality. For recent studies of the origins of pediatrics as a profession and its divided loyalties between science and social reform, the reader should consult Peter C. English's "'Not Miniature Men and Women': Abraham Jacobi's Vision of a New Medical Specialty a Century Ago," in *Children and Health Care: Moral and Social Issues* (New York: Kluwer, 1989), 247–73, and Sydney Halpern's *American Pediatrics: The Social Dynamics of Professionalism, 1880–1980* (Berkeley: University of California Press, 1988). Two works by clinicians, *History of the American Pediatric Society, 1887–1965* (New York: McGraw Hill, 1966), by Harold Faber and Rustin McIntosh, and *History of American Pediatrics* (Boston: Little, Brown, 1979), by Thomas Cone, provide useful accounts of the progress of pediatric science and professional organizations.

On the role played by infant hospitals in the evolution of pediatrics, see Peter English, "Pediatrics and the Unwanted Child in History: Foundling Homes, Disease, and the Origins of Foster Care in New York City, 1860 to 1920," *Pediatrics* 73 (1984): 699–711; Virginia A. Metaxas Quiroga, "Female Lay Managers and Scientific Pediatrics at Nursery and Child's Hospital, 1854–1910," *Bulletin of the History of Medicine* 60 (1986): 194–208; and Kathleen W. Jones's study of L. Emmett Holt's career at New York Babies' Hospital, "Sentiment and Science: The Late-Nineteenth-Century Pediatrician as Mother's Adviser," *Journal of Social History* 17 (1983): 79–96. The rise of artificial infant feeding has been examined by Rima D. Apple, *Mothers and Medicine: A Social History of Infant Feeding, 1890–1950* (Madison: University of Wisconsin Press, 1987), and Harvey Levenstein, "'Best for Babies' or 'Preventable Infanticide'? The Controversy over Artificial Feeding of Infants in America, 1880–1920," *Journal of American History* 70 (1983): 75–94. A great deal of biographical information on pediatric leaders such as Holt, Rotch, and Henry Chapin can be gleaned from *Pediatric Profiles,* edited by Borden Smith Veeder (St. Louis: Mosby, 1957), as well as *L. Emmett Holt: Pioneer of a Children's Century* (New York: D. Appleton, 1940), by R. L. Duffus and L. Emmett Holt, Jr.

Great controversy has characterized the history of obstetrics and childbirth in the United States since the 1970s. The most balanced account at present is Judith Walzer Leavitt's *Brought to Bed: Childbearing in America, 1750–1950* (New York: Oxford University Press, 1986), which highlights the role of middle-class women as agents rather than victims in the transformation of childbirth. Many other works, most notably Judy Barrett Litoff's *American Midwives: 1860 to the Present* (Westport, Conn.: Greenwood Press, 1978), focus on the eclipse of midwives. More work is needed on obstetric professionalization and therapy; Harold Speert's *Obstetrics and Gynecology in America: A History* (Baltimore: Waverly Press, 1980) offers the most helpful survey at present. The career of Joseph DeLee, recounted with the help of his nephew Sol Theron DeLee in Morris Fishbein's *Joseph Bolivar DeLee: Crusading Obstetrician* (New York: E. P. Dutton, 1949), has also recently been analyzed by Judith Walzer Leavitt in "Joseph B. DeLee and the Practice of Preventive Obstetrics," *American Journal of Public Health* 78 (1988): 1353–59.

A number of historians of women have addressed issues relevant to the treatment and rearing of infants by mothers and nurses. Sylvia Hoffert's *Private Matters: American Attitudes toward Childbearing and Infant Nurture in the Urban North, 1800–1860* (Urbana: University of Illinois Press, 1989) and Sally G. McMillen's *Motherhood in the Old South: Pregnancy, Childbirth, and Infant Rearing* (Baton Rouge: Louisiana State University Press, 1990) have surveyed the practice of mothering in the North and South, respectively, before the Civil War. More research is needed on the late nineteenth and early twentieth centuries; a useful starting point is Molly Ladd-Taylor's *Raising a Baby the Government Way: Mother's Letters to the Children's Bureau, 1915–1932* (New Brunswick, N.J.: Rutgers University Press, 1986), the best introduction to the rich resources of the U.S. Children's Bureau correspondence in the National Archives. The important role played by women physicians such as Sara Josephine Baker in early pediatric medicine and infant public health is illustrated by Regina Morantz-Sanchez in her excellent *Sympathy and Science: Women Physicians in American Medicine* (New York: Oxford University Press, 1985). Susan Reverby offers an especially provocative analysis of professional nursing in *Ordered to Care: The Dilemma of American Nursing, 1850–1945* (Cambridge: Cambridge University Press, 1987). Finally, the public role of women's voluntary organizations in shaping maternal and infant-welfare programs is brought out in Ann Firor Scott's outstanding *Natural Allies: Women's Associations in American History* (Urbana: University of Illinois Press, 1992).

The French maternal and infant-welfare movement has received considerably more attention than French medicine in the late nineteenth century. Alisa Klaus's thoroughly researched *Every Child a Lion: The Origins of Maternal and Infant Health Policy in the United States and France, 1890–1920* (Ithaca: Cornell University Press, 1993) compares the French and American infant-welfare movements. Rachel Fuchs has addressed French attitudes toward mothers and infants respectively in *Poor and Pregnant in Paris: Strategies for Survival in the Nineteenth Century* (New Brunswick, N.J.: Rutgers University Press, 1992) and *Abandoned Children: Foundlings and Child Welfare in Nineteenth-Century France* (Albany: State University of New York

Press, 1984). George Sussman's fine account of French wet-nursing practices, *Selling Mother's Milk: The Wet-Nursing Business in France, 1715–1914* (Urbana: University of Illinois Press, 1982), contains much material relevant to infant welfare.

Jack Ellis addresses the prominent role of physicians in French political life in *The Physician-Legislators of France: Medicine and Politics in the Early Third Republic, 1870–1914* (Cambridge: Cambridge University Press, 1990). Articles particularly useful for understanding the social context of French obstetricians such as Pierre Budin and Adolphe Pinard include Mary Lynn McDougall, "Protecting Infants: The French Campaign for Maternity Leaves, 1890s–1913," *French Historical Studies* 13 (1983): 79–105; Karen Offen, "Depopulation, Nationalism, and Feminism in Fin-de-Siecle France," *American Historical Review* 89 (1984): 648–76; and William Schneider, "Toward the Improvement of the Human Race: The History of Eugenics in France," *Journal of Modern History* 54 (1982): 268–91. For a comparative European case study of infant welfare, see Deborah Dwork's *War is Good for Babies and Other Children: A History of the Infant and Child Welfare Movement in England, 1898–1918* (London: Tavistock, 1987).

The historiography of medical technology remains in its infancy, with most studies thus far having centered on diagnostic instruments. Stanley Joel Reiser's *Medicine and the Reign of Technology* (Cambridge: Cambridge University Press, 1978) discusses the role played by technology in transforming medical practice, particularly with regard to supplanting the patient's experience of disease. In *Medicine and Its Technology: An Introduction to the History of Medicine* (Westport Conn.: Greenwood Press, 1982), Audrey Davis recounts the technical evolution of a number of medical tools and instruments, generally before the twentieth century. For more detailed case studies, the reader should consult Joel Howell's analyses of the X-ray and electrocardiogram, "Diagnostic Technologies: X-Rays, Electrocardiograms, and CAT Scans, *Southern California Law Review* 65 (1991): 529–64, as well as "Early Perceptions of the Electrocardiograph: From Arrythmia to Infarction," *Bulletin of the History of Medicine* 58 (1984): 83–98. Christopher Lawrence, like Howell, also emphasizes the role of social context in shaping technology in his "Incommunicable Knowledge: Science, Technology, and the Clinical Art in Britain, 1850–1914," *Journal of Contemporary History* 20 (1985): 503–20. John Harley Warner has addressed diagnostic technology in his insightful examinations of the relationship between science and medical practice, "Ideals of Science and their Discontents in Late-Nineteenth-Century American Medicine," *Isis* 82 (1991): 454–78; and "Science in Medicine," *Osiris*, 2d ser., 1 (1985): 37–58.

Other themes that intertwine closely with that of medical technology are those of medical therapeutics, the rise of the modern hospital, and medical ethics. With regard to the first of these, one should consult *The Therapeutic Revolution: Essays in the Social History of American Medicine,* ed. Morris Vogel and Charles Rosenberg (Philadelphia: University of Pennsylvania Press, 1979), as well as John Warner's *The Therapeutic Perspective: Medical Practice, Knowledge, and Identity in America, 1820–1885* (Cambridge: Harvard University Press, 1986). The two most important works for understanding the history of the American hospital are Charles Rosenberg's *The Care of Strangers: The Rise of America's Hospital System* (New York: Basic Books, 1987) and Rosemary Stevens's *In Sickness and in Wealth: American Hospitals in the Twentieth Century* (New York: Basic Books, 1989). David Rothman's *Strangers at the Bedside: A History of How Law and Bioethics Transformed Medical Decision Making* (New York: Basic Books, 1991) deals with the role of medical technology as a focus for ethical controversy.

Historians of medicine wishing to delve into the historiography of technology should begin with the essays in *The Social Construction of Technological Systems: New Directions in the Sociology and History of Technology,* ed. Wiebe Bijker, Thomas P. Hughes, and Trevor J. Pinch (Cambridge: MIT Press, 1987), along with *The Social Shaping of Technology: How the Refrigerator Got its Hum,* ed. Donald MacKenzie and Judy Wajcman (Philadelphia: Open University Press, 1985). Important case studies relating technology to social context include Merritt Roe Smith, *Harpers Ferry Armory and the New Technology: The Challenge of Change* (Ithaca: Cornell University Press, 1977); Thomas P. Hughes, *Networks of Power: Electrification in Western Society, 1880–1930* (Baltimore: Johns Hopkins University Press, 1983); and Donald Mackenzie, *Inventing Accuracy: An Historical Sociology of Nuclear Missle Guidance* (Cambridge: MIT Press, 1990). Ruth Schwartz Cowan provides a fascinating analysis of domestic technology in *More Work for Mother: The Ironies of Household Technology from the Open Hearth to the Microwave* (New York: Basic Books, 1983). Finally, for a historical overview of technology in the United States see Thomas Hughes' *American Genesis: A Century of Invention and Technological Enthusiasm, 1870–1970* (New York: Viking Press, 1989).

INDEX